Letts

Physics

Gurinder Chadha

Contents

A2 exams

Different types of questions

In A2 examinations, there are mainly two sorts of examination questions: ***structured*** questions and ***extended writing*** questions.

Structured questions

In a structured question, a collection of common ideas is tested and the question is set out in smaller sub-sections. The earlier sections of the questions have a tendency to be easier and are designed to ease you into the question. Sometimes the answer to a particular section is given, so as to point you in the right direction. The later sections of a question can be a little discriminatory and you might have to think very carefully about your response. The following guide may be helpful:

- Read each question with great care and underline or highlight any key terms or data given.
- Write your answers carefully and always show **all** stages of your calculations.
- At every opportunity, keep referring to any introductory section to the question.

Extended questions

In an extended question, examiners will be assessing your understanding of physics and your ability to communicate ideas effectively and clearly. It is vital that in such questions, you pay particular attention to spellings, grammar and sentence construction. The response to such questions can be open-ended, with examiners keen to award marks for any 'good', but relevant, physics. The following guide may be helpful:

- Read the question carefully and make sure that you do clearly understand what the examiners want.
- Make a quick and short plan of your key ideas. It is worth putting your ideas in a ***hierarchical*** order.
- Write your answers clearly and remember to pay particular attention to the way you write.

Other questions

In addition to structured and extended questions, examiners also make use of free-response and open-ended questions. The points raised in the guide above are equally applicable here.

What examiners look for

An examination paper is devised to assess your knowledge, understanding and application of physics. One of the main purposes of an examination paper is to sort out how good you are at physics. Examiners are looking for the following points:

- Correct answers to the questions.
- Clearly presented answers with all the working shown.
- Concise answers to structured questions and logically set out responses to extended questions.
- Sketch graphs and diagrams that are drawn neatly, with particular attention given to labels and units.

What makes an A, C and E candidate?

The examiners can only ask questions that are within the specifications (this is the new term for the 'syllabus'). It is therefore vital that you are fully aware of what the examiners can and cannot ask in any examination. Be prepared to tackle the complexities of the paper. Your aim must be to achieve high marks in unit or module examinations. The way to accomplish this is to have a good knowledge and understanding of physics. Listed below are the minimum marks for grades A to E.

Grade A 80% **Grade B** 70% **Grade C** 60% **Grade D** 50% **Grade E** 40%

- **A grade candidates** have an excellent all-round knowledge of physics and they can apply that knowledge to new situations. Such candidates tend to be strong in all of the modules and tend to have excellent recall skills.
- **C grade candidates** have a reasonable knowledge of physics but have some problems when they apply their knowledge to new situations. They have some gaps in their knowledge and tend to be weak in some of the modules.
- **E grade candidates** have a poor knowledge of physics and have not learnt to apply their ideas to familiar and new situations. Such candidates find it difficult to recall key definitions and equations.

Successful revision

Revision skills

- Always start with a topic that you find easier. This will boost your self-confidence.
- Do not revise for too long. When you are tired and irritable, you cannot produce quality work.
- Make notes on post cards or lined paper of key ideas and equations. Do not feel that you have to write down everything. Just the key points need to be jotted down. Sometimes you have to learn certain proofs. It is worth writing down all the important steps for such proofs.
- Make good use of the specification. Use a highlighter pen to identify topics that you have already revised.
- Do not leave your revision to the last moment. Plan out a strategy spread over many weeks before the actual examination. Work hard during the day and learn to relax when needed.
- Whatever happens, do not try to learn any new topics on the day before the examination. It is important for you to be calm and relaxed for the actual examination.

Practice questions

This book is designed to improve your understanding of physics and, of course, improve your final grade.

Look carefully at the grade A and C candidates' responses. Can you do better? There are some important tips given to improve your understanding.

Try the practice examination questions and then look at the answers and tips given.

When you are ready, try the A2 mock examination papers.

Planning and timing your answers in the exam

- Write legibly and stay focused throughout.
- Sometimes, candidates think that they have answered all the questions and then find an entire question on the last page. You do not want to be in this predicament, so **quickly** scan through the entire paper to see what you have to do.
- Do the question on the paper that you are most comfortable with. This will boost your confidence.
- Read each question carefully. Highlight the key ideas and data. The information given is there to be used.
- As a very rough guide, you have about 1 minute for each mark. The number of lines allocated for your answer gives you an idea of the depth and detail required for a particular answer. The marks allocated for each sub-section give an idea of how many steps or items of information are required.
- Do a quick plan for extended questions. It is not sensible to start writing straight away because you will end up either repeating yourself or missing out some important ideas.
- Do not use correction fluid. If you have strong reasons that a particular answer is wrong, then simply cross it out and provide an alternative answer.

Setting out numerical answers

It is important that your answers to numerical and algebraic questions are set out logically for the examiner. In this book, a simple method is used to indicate **where** a mark is awarded for the correct response. This is indicated by means of a tick (✔).

The **marking scheme** adopted in this book and how you ought to structure a numerical answer is illustrated in the example below.

Question: What is the pressure exerted by a force of 9.0 kN acting on an area of 1.5×10^{-2} m^2? **[2]**

Answers: $P = F/A$ ✔ (Make the physics clear to the examiner.)

$P = 9.0 \times 10^3/1.5 \times 10^{-2}$ (Use standard form and remember to convert $k \rightarrow 10^3$.)

$P = 6.0 \times 10^5$ Pa ✔ (Do not forget the correct unit and significant figures.)

There are only two marks for the calculation. One mark is awarded for the equation and the other for the correct answer and the unit.

Remember, the ticks appear next to responses where the marks are awarded.

No credit can be given for a **wrong** answer. However, by writing down all the stages of your work, it may be possible to pick up some or all of the 'part marks'. So help yourself and set out your work in a clear and methodical way.

How to boost your grade

What examiners look for

Examiners cannot give credit for the wrong physics. A wrongly quoted equation cannot be awarded any marks. For an examiner, it is quite disturbing to find candidates who cannot re-arrange equations. Mathematics is the language of physics, therefore it is important for candidates to be comfortable with handling and re-arranging equations. There are several techniques for re-arranging equations, but the one outlined below can be learnt and applied quickly.

Remember BODMAS.

When solving a numerical or algebraic equation, you must do the mathematical operations in the order given by the mnemonic BODMAS.

Bracket → **O**f → **D**ivision → **M**ultiplication → **A**ddition → **S**ubtraction

When it comes to re-arranging an equation, you simply **reverse** the mathematical operations. Here is an example to illustrate this technique for re-arranging an equation.

$v^2 = u^2 + 2as$ What is a?

Using the ideas developed above, we have

$$a \rightarrow (\times 2s) \rightarrow (+ u^2) \rightarrow = v^2$$

By reversing the sequence and carrying out the inverse operations, we end up with

$$v^2 \rightarrow (- u^2) \rightarrow (\div 2s) \rightarrow = a$$

Therefore $$a = \frac{(v^2 - u^2)}{2s}$$

If you have some other tried and tested technique for re-arranging equations, then it is best to stick to it. However, do remember to take re-arranging of equations seriously in physics.

Here are some other suggestions to **boost your grade**.

- Be familiar with the specifications.
- Learn all the definitions within the specifications. Recalling definitions can give you easy marks and improve your final grade.
- Write all the stages of a numerical solution. If your final answer is wrong, you still have a chance to pick up some of the 'part-marks'.
- In a question with 'state', the answer is brief and does not require any further explanation. In a question with 'describe', the answer can be long and may require full explanation of some physics.
- Use the information given in the question to guide your answers. For numerical solutions, keep an eye on the significant figures and units. Your final answer must not be more or less than the significant figures given in the question. In physics, it is sensible to write the final answer in standard form, e.g. 1.62×10^{-3} A.
- Read the information given on graphs and tables carefully. Sometimes data is given in either standard or prefix form. Do not forget to take this on board when doing your calculations. For example, the stress axis is labelled as 'stress/MPa'. Remember that 'MPa' is 10^6 Pa.
- It is easy to press the wrong buttons on the calculator. Make sure that your answer looks reasonable. If you have time, you ought to check a calculation again.
- Draw diagrams carefully and make sure that you label all the key items.
- Your graphs must be correctly labelled and have a suitable scale so that they fill most of the graph paper.
- You do not have to recall physical data. All the data required is normally given on the question paper itself or on a separate data-sheet.

For more information about your course go to:
www.aqa.org.uk
www.ccea.org.uk
www.edexcel.org.uk
www.wjec.org.uk
www.ocr.org.uk

Chapter 1 Mechanics

Questions with model answers

C grade candidate – mark scored 6/10

1 A jet of water from a hose pipe hits a wall at right angles to it and then trickles down the wall.

(a) Explain how the water jet exerts a force on the wall. [3]

The water jet has mass and velocity and therefore momentum. ✔
It hits the wall with a force. ✗
The jet loses its power and then trickles down the wall. ✗

(b) Calculate the magnitude of the force exerted on the wall by the water jet if the water is delivered at a rate of 8.0 kg s^{-1} and hits the wall with a velocity of 25 m s^{-1}. [2]

$F = \frac{m(v-u)}{t}$ ✔

$F = \left(\frac{m}{t}\right) \times u$ (magnitude only)

$F = 8.0 \times 25 = 200$ N ✔

2 A tennis player hits an incoming ball of mass 60 g with a racquet and changes its direction of travel. The force exerted by the racquet on the ball is 150 N. The diagram shows how the velocity of the ball is changed by the racquet.

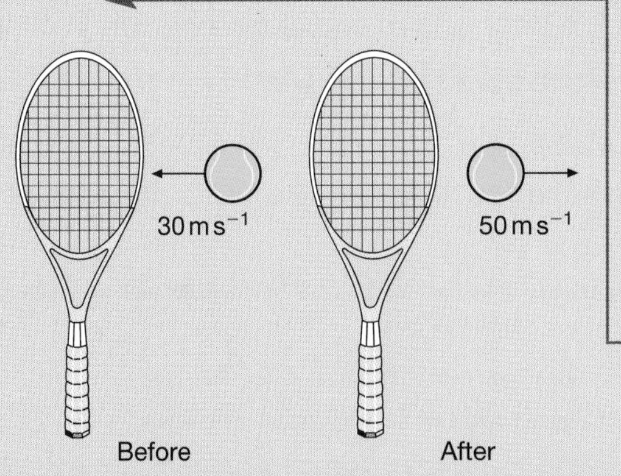

(a) Calculate the magnitude of the impulse exerted by the racquet. [2]

Impulse = $F\Delta t$
Impulse = change in momentum ✔
Impulse = $(0.06 \times 50) - (0.06 \times 30)$ ✗
Impulse = 1.2 N s

(b) Hence calculate the time of impact of the ball with the racquet. [3]

Impulse = change in momentum of ball
$F\Delta t = 1.2$ ✔
$\Delta t = \frac{1.2}{150}$ ✔
$\Delta t = 8.0 \times 10^{-3}$ ✗

error carried forward

GRADE BOOSTER

To improve the grade, this candidate should have read the question carefully. In question 2, the candidate lost two marks for not knowing the vector nature of momentum and missing out the unit for Δt. An extra two marks would have made the answer an A grade.

Examiner's Commentary

For help see Revise A2 Study Guide sections 1.1 and 1.2

This is a disappointing answer. One mark has been awarded for the first statement, but the other two statements are either incomplete or misleading. On impact with the wall, there is a **change** in momentum for the water jet. According to Newton's second law, a change in momentum in a given time results in a force on the water jet. This force is provided by the wall. The water jet exerts an equal but opposite force on the wall (Newton's third law). The magnitude of the force depends on the velocity of the water jet and on the rate at which the mass of water hits the wall.

The velocity of the water after impact is zero. Hence, the candidate has correctly made the final velocity 'v' equal to zero.

Momentum is a vector quantity and as such, it has both magnitude and direction. The momentum of the ball after impact with the racquet has to be of the opposite sign to the initial momentum. The magnitude of the impulse provided by the racquet is:

impulse = $(0.06 \times 50) - (0.06 \times -30)$
impulse = 4.8 kg m s^{-1}

The candidate has used the wrong answer of 1.2 N s from **(a)**. This mistake has already been penalised. The examiner has been fair and awarded all subsequent marks. Sadly, the candidate lost the final mark because the unit for time has been missed out.

A grade candidate – mark scored 8/8

1 Explain how the tyres of a car generate a forward force on the car. [3]

The car tyres rotate ✔ in such as way that they exert a backward force on the ground. ✔
According to Newton's third law, 'action = reaction', therefore the ground exerts an equal forward force on the tyres. ✔

2 A car and its occupants of total mass of 1.2×10^3 kg is moving along the motorway. At a particular time, its speed and acceleration are $15\ \text{m s}^{-1}$ and $0.2\ \text{m s}^{-2}$ respectively. At this time the total resistive force on the car is 150 N.

(a) Name one of the resistive forces on the car. [1]

As the car moves through the air, the air exerts a resistive force called drag. ✔

(b) Calculate the resultant force on the car. [2]

$F = ma$ ✔
$F = 1.2 \times 10^3 \times 0.2$
$F = 240$ N ✔

(c) Calculate the instantaneous power developed by the engine of the car. [2]

$P = Fv$
$F = 150 + 240 = 390$ N ✔
Power $= 390 \times 15 = 5.9$ kW ✔

Examiner's Commentary

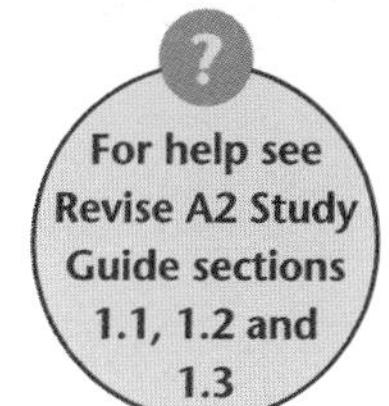

For help see Revise A2 Study Guide sections 1.1, 1.2 and 1.3

The motive force provided by the car is 390 N. In order to determine the power developed by the car, it would be incorrect to use just the resultant force from **(b)**.

Exam practice questions

1 **(a)** The diagram shows a rocket travelling in space.

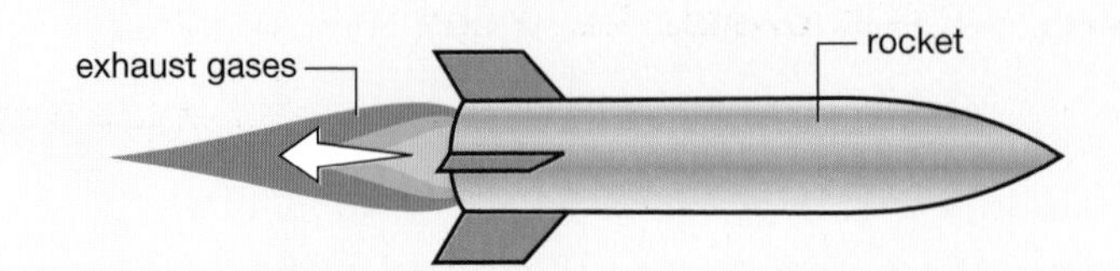

Use your knowledge of Newton's laws to explain the origin of the force on the rocket as it expels exhaust gases at high velocity. [4]

(b) A bullet of mass 1.40×10^{-2} kg is fired horizontally from a gun with a velocity of 210 m s^{-1}. It hits and gets embedded within a stationary wooden block. The block of wood has a mass of 1.50 kg and lies on a horizontal frictionless surface. After impact, the wooden block (together with the embedded bullet) moves with a constant velocity.

(i) Calculate the momentum of the bullet just before it enters the wooden block. [2]

(ii) Calculate the velocity, v, of the wooden block after being hit by the bullet. [3]

(iii) Explain whether or not kinetic energy is conserved in the impact. [2]

[Total: 11]

---

2 The diagram shows a 70 kg athlete running up a hill that is sloping at an angle of 6.0° to the horizontal. The athlete maintains a constant speed whilst running up the hill.

In 50 s, the athlete covers 200 m along the hill. Calculate

(a) the vertical distance h climbed by the athlete, [2]

(b) the gravitational potential energy gained by the athlete,
Data: $g = 9.8$ N kg^{-1} [2]

(c) the power developed by the athlete. [2]

[Total: 6]

Answers on pages 14–19 Answers on pages 14–19 Answers on pages 14–19

3 **(a)** Define work done by a force. **[2]**

(b) An object is **released** close to the surface of the Earth.

(i) Use the idea of work done by a force, to explain why the speed of the object increases as it falls towards the ground. **[3]**

(ii) Explain whether or not momentum is conserved as the object accelerates towards the ground. **[2]**

(c) A swing consists of a rubber tyre of mass 14 kg suspended from a 6.0 m long rope.
A child of mass 35 kg sits on the swing. The speed of the child at the lowest point is $4.0\,\mathrm{m\,s^{-1}}$.

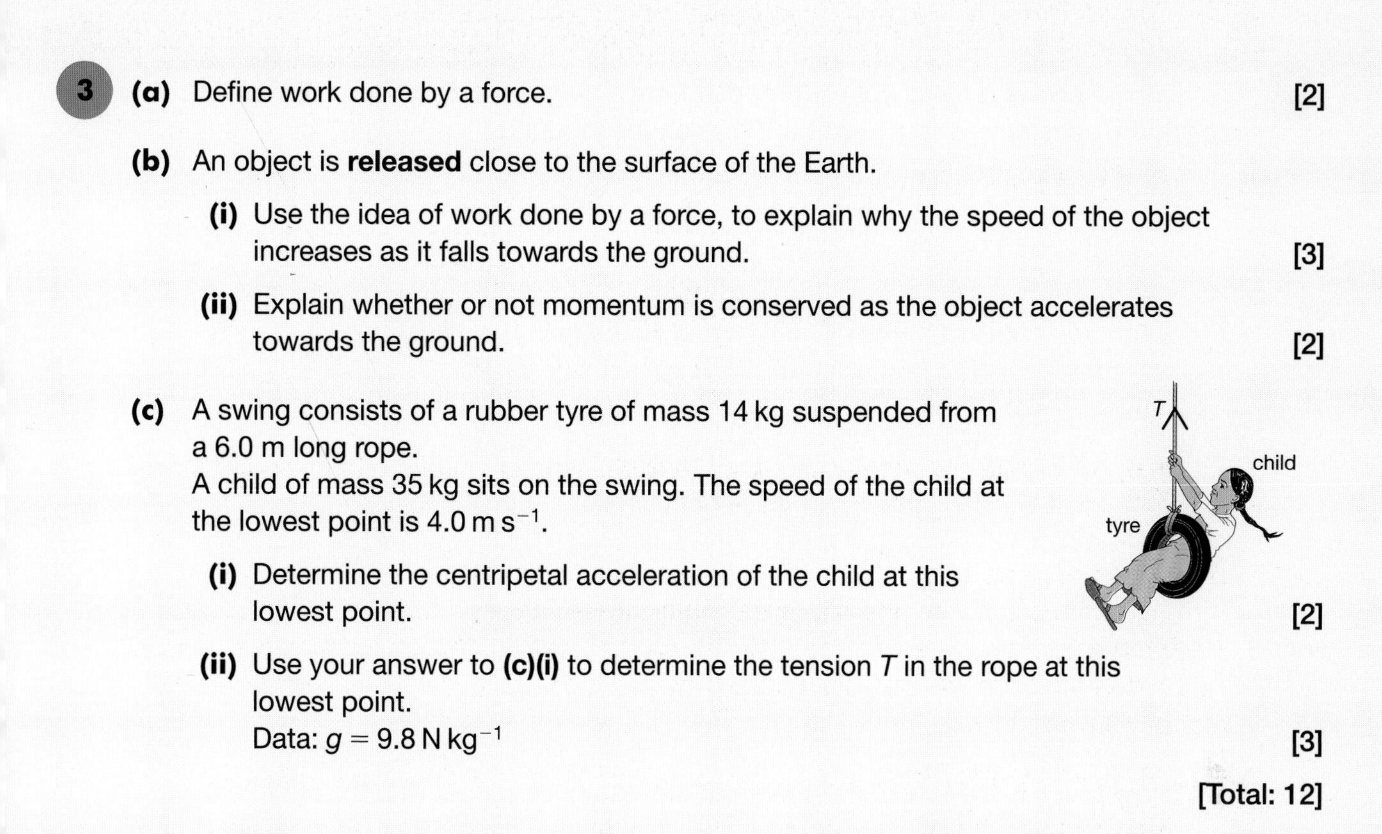

(i) Determine the centripetal acceleration of the child at this lowest point. **[2]**

(ii) Use your answer to **(c)(i)** to determine the tension T in the rope at this lowest point.
Data: $g = 9.8\,\mathrm{N\,kg^{-1}}$ **[3]**

[Total: 12]

4 **(a)** Define momentum and state its unit. **[2]**

(b) A stationary nucleus of thorium-226 decays by alpha-particle emission into radium. The decay equation is:

$$^{226}_{90}\mathrm{Th} \rightarrow {}^{222}_{88}\mathrm{Ra} + {}^{4}_{2}\mathrm{He}$$

(i) State the value of the momentum of the thorium nucleus before the decay. **[1]**

(ii) After the decay, both the alpha-particle and the radium nucleus are moving. Which has the greater speed? Justify your answer. **[2]**

(iii) What can be said about the directions of travel of the two particles? **[1]**

[Edexcel June 2003]

[Total: 6]

5 The diagram shows a metal sphere dropped from a height of 5.6 m onto a flat surface. After impact, the sphere rises to a height of 1.2 m on the rebound.

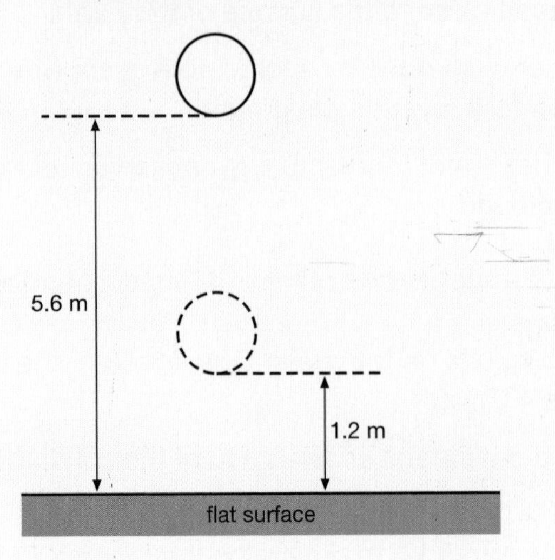

Data: $g = 9.8\ \text{m s}^{-2}$

(a) Show that the speed v of the sphere at the instant of impact with the surface is given by

$$v = \sqrt{2gH}$$

where g is the acceleration of free fall and H is the vertical distance through which the sphere drops. [3]

(b) Calculate the speed of the sphere at the instant of impact. [1]

(c) Calculate the time taken for the sphere to fall through the vertical distance of 5.6 m. [3]

(d) Calculate the magnitude of the force exerted on the sphere by the surface, given that the duration of the impact is 75 ms. The mass of the sphere is 260 g. [3]

(e) Sketch a graph of velocity against time from the instant the sphere is dropped to the time that it rebounds to the height of 1.2 m. [3]

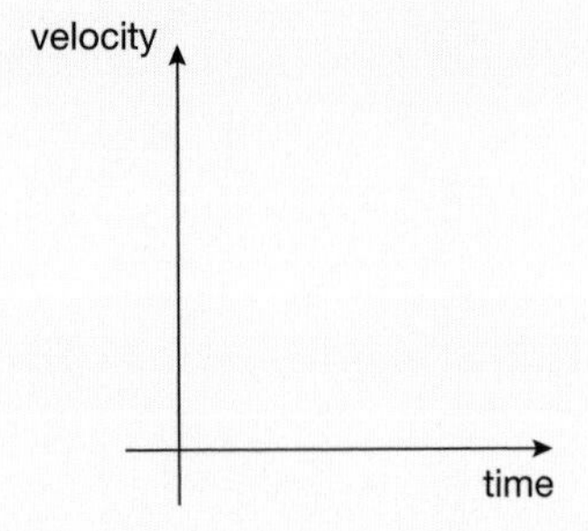

[Total: 13]

Mechanics

Answers on pages 14–19 Answers on pages 14–19 Answers on pages 14–19

6 The diagram shows a toy car of mass 80 g gently dropped onto a smooth circular track of radius 32 cm.

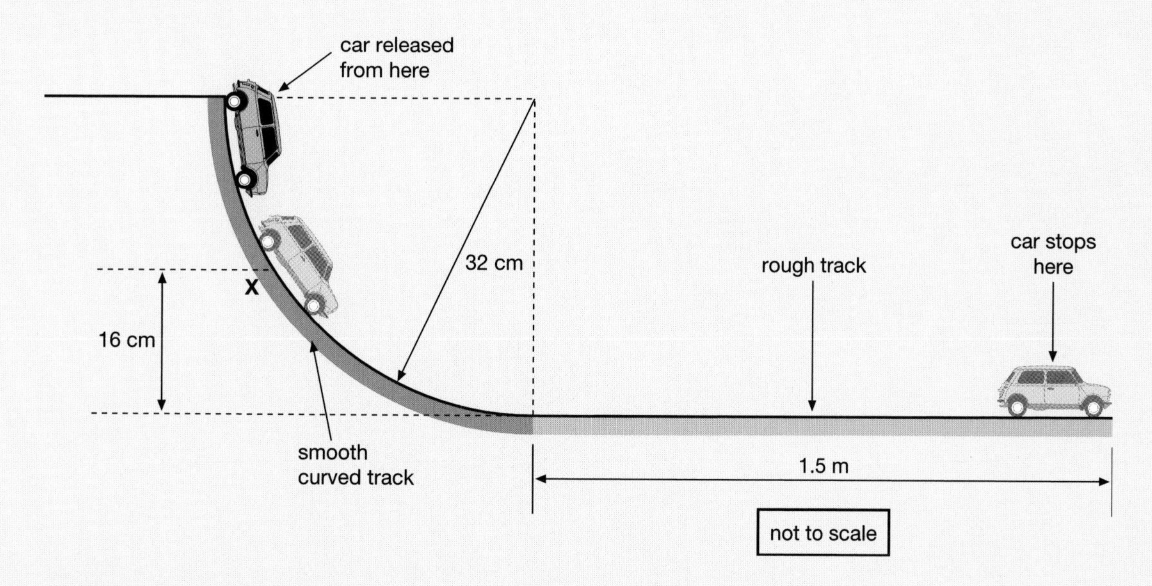

(a) Show that the speed of the car at point **X** is 1.8 m s^{-1}. [3]
Data: $g = 9.8$ m s^{-2}

(b) Calculate the centripetal force acting on the car when at point **X**. [3]

(c) After reaching the bottom of the curved track, the car travels along a rough horizontal track and stops after travelling a distance of 1.5 m. Calculate the average friction acting on the car. [3]

[Total: 9]

7 The diagram shows a pendulum bob of mass 120 g which has been set to move in a horizontal circle at a constant speed.

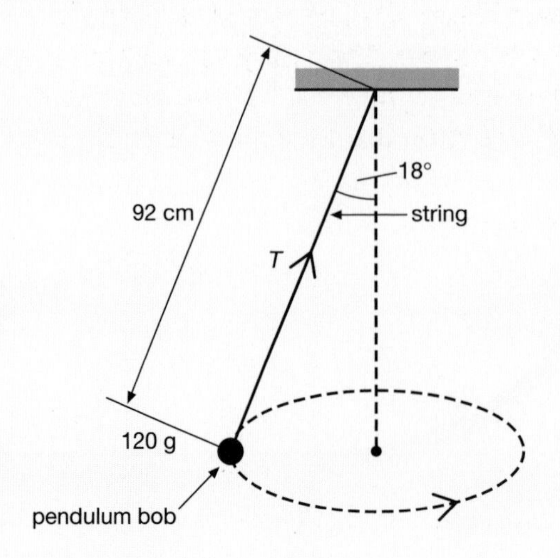

Data: $g = 9.8\ \text{m s}^{-2}$

(a) Show that the tension T in the string is about 1.2 N. [3]

(b) Calculate the speed v of the pendulum bob. [3]

[Total: 6]

---

8 **(a)** State Newton's third law. [2]

(b) State the principle of conservation of momentum. [2]

(c) The diagram shows two rugby players about to collide head-on.

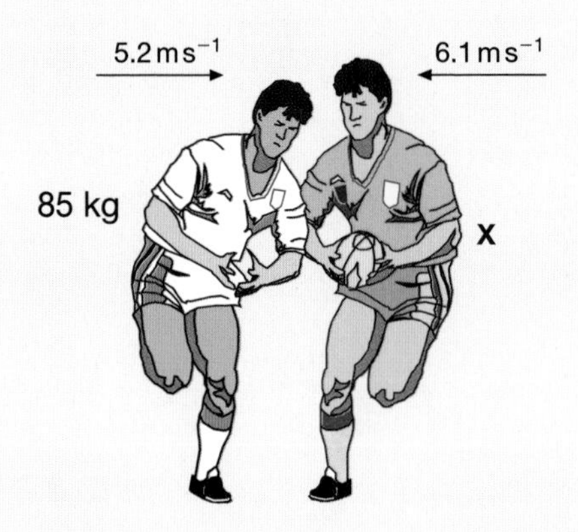

Immediately after the collision, the rugby players get tangled and have a common velocity of $0.60\ \text{m s}^{-1}$ towards the left.

(i) Calculate the mass of the rugby player marked **X**. [3]

(ii) Explain, with the aid of a calculation, whether or not the collision is elastic. [3]

[Total: 10]

Mechanics

Answers on pages 14–19 **Answers** on pages 14–19 **Answers** on pages 14–19

(a) Define the impulse of a force. **[1]**

(b) A racing car of mass 750 kg travelling at a steady speed crashes into a safety barrier. During the collision, the car and the safety barrier both buckle. The car stops moving in its original direction after a time of 1.2 s. The diagram below shows the variation of the retarding force *F* acting on the car with time *t* from the moment of the initial impact.

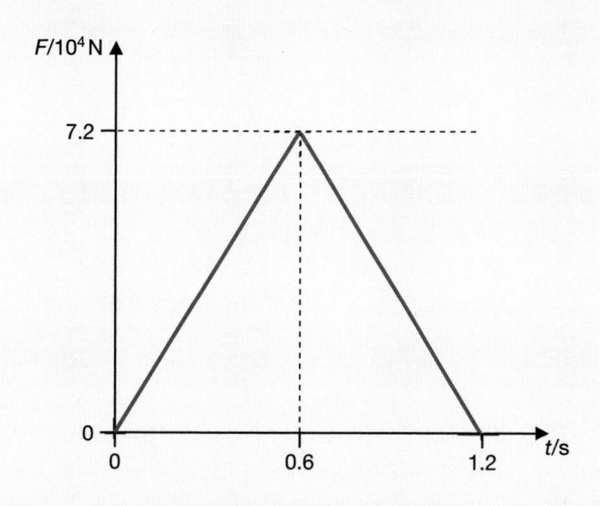

(i) Calculate the impulse required to stop the car. **[2]**

(ii) Calculate the initial momentum of the car. **[2]**

(iii) Calculate the speed of the car just before making its impact with the safety barrier. **[2]**

[Total: 7]

Mechanics

Answers on pages 14–19 Answers on pages 14–19 Answers on pages 14–19

Answers

(1) (a) The exhaust gases have momentum.
A rate of change of momentum takes place for the exhaust gases.
According to Newton's second law, there is a force exerted on the exhaust gases.
According to Newton's third law, the force acting **on** the exhaust gases is equal and opposite to that experienced **by** the rocket.
Hence there is a force acting on the rocket.

(b) (i) $p = mv$
$p = 1.40 \times 10^{-2} \times 210$
$p = 2.94\ \text{kg m s}^{-1}$

(ii) Total mass after impact $= M$
$M = 1.50 + 0.014 = 1.514\ \text{kg}$
Momentum is conserved in the 'collision'.
$\therefore 2.94 = 1.514\ v$
$v = 1.94\ \text{m s}^{-1}$

(iii) Kinetic energy of the bullet is **not** conserved.
Some of the kinetic energy of the bullet is converted into heat, sound, etc.
The impact is an inelastic collision.

examiner's tip

In all collisions, where there are no external forces acting, the momentum is always conserved. Hence

Initial momentum = final momentum

In this collision, energy is also conserved. However, kinetic energy is not. Some of the kinetic energy of the bullet is used to make a dent in the wood, generate sound, etc. So, it would be wrong to suggest that kinetic energy is conserved in this impact. You cannot therefore write the following:

initial K.E. = final K.E.

$$\tfrac{1}{2} \times 1.40 \times 10^{-2} \times 210^2 = \tfrac{1}{2} \times 1.514 \times v^2.$$

---

(2) (a) $\sin 6.0° = \dfrac{h}{200}$

$h = 200 \times \sin 6.0°$
$\therefore h = 20.9 \approx 21\ \text{m}$

(b) Gain in gravitational potential energy = G.P.E.
G.P.E. $= mg\Delta h$
G.P.E. $= 70 \times 9.8 \times 20.9$
G.P.E. $= 1.43 \times 10^4 \approx 1.4 \times 10^4\ \text{J}$

(c) $\text{Power} = \dfrac{\text{energy transfer}}{\text{time taken}}$

$P = \dfrac{1.43 \times 10^4}{50}$

$P = 286 \approx 290\ \text{W}$

Mechanics

(3) (a) Work done is defined as the product of the force and the distance moved by the force.
The distance moved is that in the **direction** of the force.

(b) (i) The weight of the object accelerates it towards the Earth.
Work is done on the object by the weight.
Work done by the weight = gain in kinetic energy.
Therefore the speed of the falling object increases during its descent.

(ii) The falling object exerts an equal but opposite force on the Earth.
The Earth acquires an **equal** but **opposite** momentum to the falling object.
The total momentum is zero.
Momentum is therefore conserved.

examiner's tip

It is very important to appreciate the idea that momentum is a vector quantity.

(c) (i) $a = \frac{v^2}{r}$

$a = \frac{4.0^2}{6.0}$

$a = 2.67 \approx 2.7\ \text{m s}^{-2}$

(ii) m = total mass

$m = 35 + 14 = 49\ \text{kg}$

$T - mg = ma$

$T = ma + mg$

$T = 49 \times (2.67 + 9.8)$

$T = 611 \approx 610\ \text{N}$

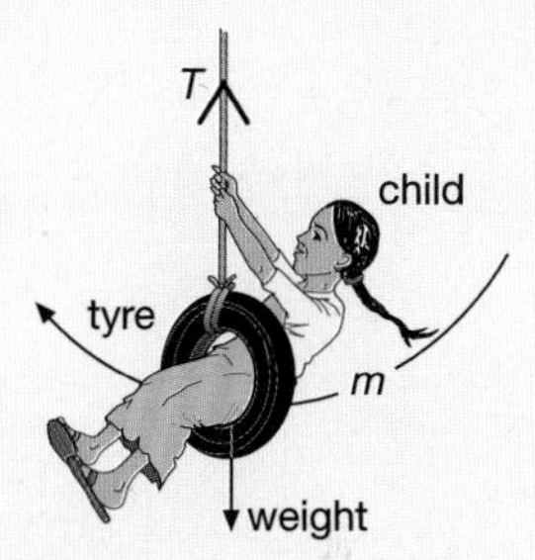

examiner's tip

It is very tempting to calculate the tension T in the rope by using

$$T = ma.$$

Such an approach would be wrong. The resultant force at the lowest point of the swing is

$$T - \text{weight}.$$

Once the resultant force has been identified, Newton's second law may be used to calculate the tension T in the rope. Therefore

$$T - mg = ma$$

Mechanics

(4) (a) momentum = mass × velocity
unit → [kg m s^{-1}]

> **examiner's tip**
> **You must remember to use 'velocity' in the definition and not 'speed'.**

(b) (i) Zero

(ii) The particles have equal and opposite momemtum because momentum is conserved. The alpha-particle has the greater speed because it has a smaller mass than the radium nucleus.

(iii) They travel in opposite directions.

> **examiner's tip**
> **Momentum is a vector quantity.**
> **The radium-222 nucleus and the alpha-particle travel in opposite directions; each particle has momentum of equal magnitude. The radium nucleus is approximately $\frac{222}{4} \approx 56$ times more massive than the alpha-particle. Hence the speed of the alpha-particle will be approximately 56 times greater than that of the radium nucleus.**

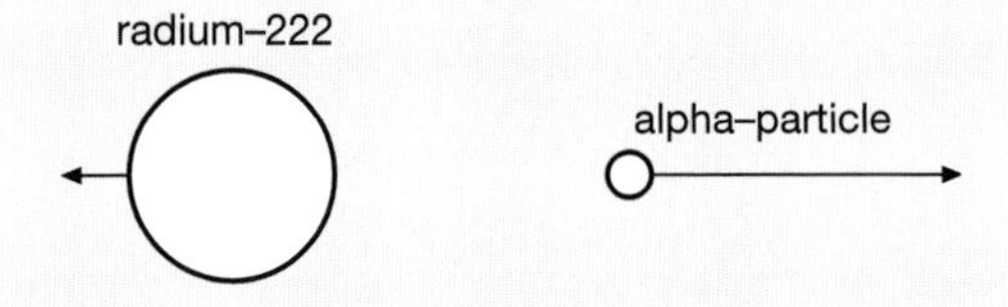

---

(5) (a) loss in gravitational potential energy = gain in kinetic energy

$$mgH = \frac{1}{2}mv^2$$

The mass on either side of the equation cancels, therefore

$$gH = \frac{v^2}{2}$$

$$v^2 = 2gH$$

$$v = \sqrt{2gH}$$

> **examiner's tip**
> **This is an important proof because it shows that the mass of the object is not a factor in determining the final speed of the object falling in the Earth's gravitational field. Learn this proof.**

(b) $v = \sqrt{2gH} = \sqrt{2 \times 9.8 \times 5.6}$

$v = 10.48 \text{ m s}^{-1} \approx 10.5 \text{ m s}^{-1}$

(c) $s = ut + \frac{1}{2}at^2 \qquad (u = 0)$

$$t = \sqrt{\frac{2s}{a}} = \sqrt{\frac{2 \times 5.6}{9.8}}$$

$t = 1.07\ s$

Mechanics

(d) final speed of sphere after impact $= \sqrt{2gH} = \sqrt{2 \times 9.8 \times 1.2}$

final speed $= 4.85\ \text{m s}^{-1}$

force = rate of change of momentum

$$\text{force} = \frac{[0.260 \times (-4.85)] - [0.260 \times (+10.48)]}{0.075} \approx -53\ \text{N}$$

magnitude of the force = 53 N

examiner's tip

Before you can apply Newton's second law, you have to determine the velocity of the sphere just after the impact and also remember to take the vector nature of velocity into account.
In this question, it is sensible to take the downward direction as being positive because the falling object has an acceleration of $+9.8\ \text{m s}^{-2}$. Hence, the velocity just after the impact is $-4.85\ \text{m s}^{-1}$ and velocity just before the impact is $+10.48\ \text{m s}^{-1}$.

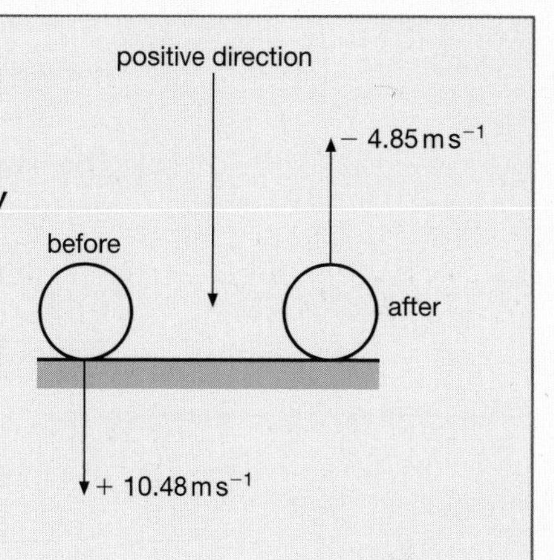

(e)

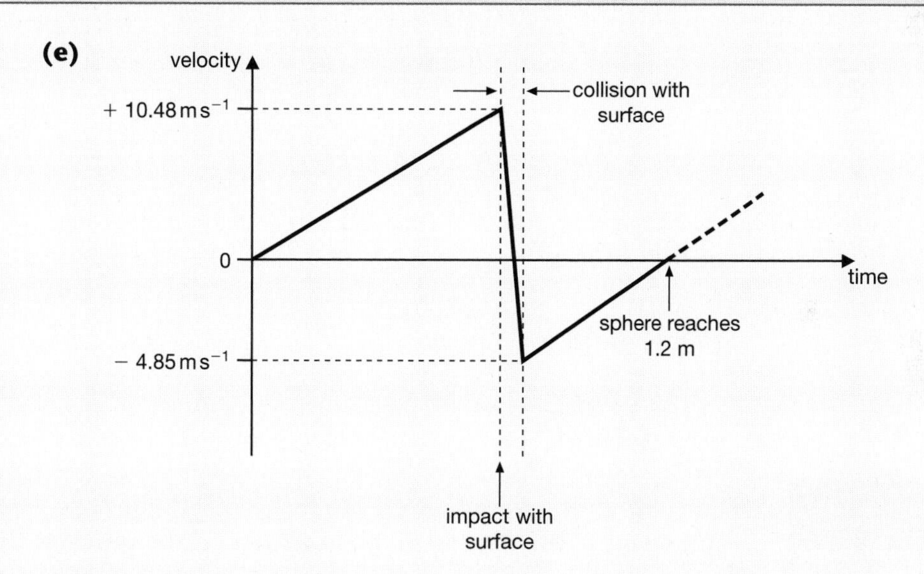

examiner's tip

The gradient from the velocity against time graph is equal to the acceleration. Therefore, during the free fall of the sphere, the graph is a straight line with a positive slope and having a gradient equal to $+9.8\ \text{m s}^{-2}$. Notice the massive deceleration during the impact with the surface. In an examination, you can also secure full marks for drawing a 'mirror image' graph because you may have taken the downward direction as being negative. As long as you are consistent, you will not lose marks.

(6) (a) gain in kinetic energy = loss in gravitational potential energy

$\frac{1}{2}mv^2 = mgh$

$v = \sqrt{2gh} = \sqrt{2 \times 9.8 \times 0.16}$ (The car falls a vertical distance of 16 cm.)

$v = 1.77\ \text{m s}^{-1} \approx 1.8\ \text{m s}^{-1}$

(b) $F = ma$

$$F = m\frac{v^2}{r} = \frac{0.080 \times 1.77^2}{0.32}$$

net force = 0.78 N

(c) The final kinetic energy at the bottom of the curved track is equal to the initial gravitational potential energy of the car.

work done against friction = kinetic energy at bottom of curved track

$F \times 1.5 = 0.080 \times 9.8 \times 0.32$

$F = 0.17\ \text{N}$

---

(7) (a) Net force in the vertical direction = zero

$T \cos 18° = mg$

$T \times 0.951 = 0.120 \times 9.8$

$$T = \frac{0.120 \times 9.8}{0.951} = 1.24\ \text{N}$$

(b) radius of circle = 92 sin 18° = 28.4 cm

centripetal force = ma

$$T \sin 18° = \frac{mv^2}{r}$$

$$1.24 \sin 18° = \frac{0.120 \times v^2}{0.284}$$

$v^2 = 0.907$

$v = 0.95\ \text{m s}^{-1}$

examiner's tip

There are several mistakes you can make in this question. The resolved component of the tension in the horizontal plane provides the centripetal force on the pendulum bob. You must also make sure the radius of the circle is correctly calculated. You would not score many marks for using the force of 1.24 N and a radius equal to 92 cm to determine the speed of the pendulum bob.

Mechanics

(8) (a) When two bodies interact, each exerts a force on the other which is equal in magnitude but opposite in direction. This is true as long as the system is 'closed' (that is, no external forces act on the objects).

(b) In a closed system, the total initial momentum of the system before the collision is equal to the total final momentum after the collision.

(c) (i) initial momentum = final momentum

$(85 \times 5.2) + (m \times -6.1) = (85 + m) \times -0.60$

$442 - 6.1m = -51 - 0.60m$

$493 = 5.5m$

$m = 89.6 \text{ kg} \approx 90 \text{ kg}$

examiner's tip

Do not forget the vector nature of both velocity and momentum. It is vital that the velocities are given the correct signs.

(ii) total initial kinetic energy $= (\frac{1}{2} \times 85 \times 5.2^2) + (\frac{1}{2} \times 89.6 \times 6.1^2) \approx 2800$ J

total final kinetic energy $= \frac{1}{2} \times (89.6 + 85) \times 0.60^2 \approx 31$ J

The collision is inelastic because the kinetic energy is not conserved.

---

(9) (a) impulse = force × time

(b) (i) impulse = area under the force against time graph

impulse $\approx \frac{1}{2} \times 7.2 \times 10^4 \times 1.2 = 4.32 \times 10^4$ N s

(ii) impulse = change in momentum of the car

final momentum = 0

Hence, initial momentum $= 4.32 \times 10^4$ kg m s^{-1}

examiner's tip

According to Newton's second law: Force = rate of change of momentum

$$F = \frac{\Delta p}{\Delta t}$$

Therefore

$$F\Delta t = \Delta p$$

impulse = change in momentum

(iii) $mv = 4.32 \times 10^4$

$$v = \frac{4.32 \times 10^4}{750} = 58 \text{ m s}^{-1}$$

Mechanics

Chapter 2 **Waves and oscillations**

Questions with model answers

C grade candidate – mark scored 5/8

Examiner's Commentary

(a) This question is about simple harmonic motion (s.h.m.). Which of the following statements are true? Place a tick against a correct response. **[2]**

	Place a tick here
A tennis ball being hit repeatedly over the net executes shm.	
A simple harmonic oscillator has maximum acceleration when its displacement is a maximum.	✔
The acceleration is always directed towards some fixed point.	✔
The period of a simple harmonic oscillator depends on its amplitude.	

✔

✔

(b) A metal strip clamped to the edge of a table has an object of mass 280 g attached to its free end. The object is pulled down and then released. The object executes simple harmonic motion with an amplitude of 8.0 cm and a period of 0.16 s.

(i) Calculate the maximum acceleration of the object. **[3]**

$a = -\omega^2 x$

$a = -(2\pi f)^2 \times 0.08$ ✔

$a = -(2\pi \times 0.16) \times 0.08 = 0.08\ \text{m s}^{-2}$ ✗

This candidate does not appreciate the difference between frequency and period. The situation is made worse by not squaring the ω term. The correct answer is

$a = -(2\pi/0.16)^2 \times 0.08 \approx 120\ \text{m s}^{-2}$

(ii) Calculate the maximum force experienced by the object. **[2]**

$F = ma$ ✔

$F = 0.280 \times 0.08 = 0.0224\ \text{N}$ ✔

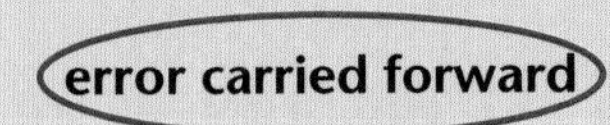

(iii) State the position of the object when it has no kinetic energy. **[1]**

The object has no kinetic energy when its displacement is zero. ✗

The velocity and kinetic energy of the object is zero when the displacement is maximum and equal to the amplitude.

GRADE BOOSTER

You cannot afford to have any gaps in your knowledge. This candidate could have got a higher grade by knowing how to calculate the frequency of the oscillator.

A grade candidate – mark scored 7/7

Examiner's Commentary

Microwaves of frequency 12 GHz are incident on a large piece of metal with a 15 cm wide slit. The microwaves are diffracted at the slit.

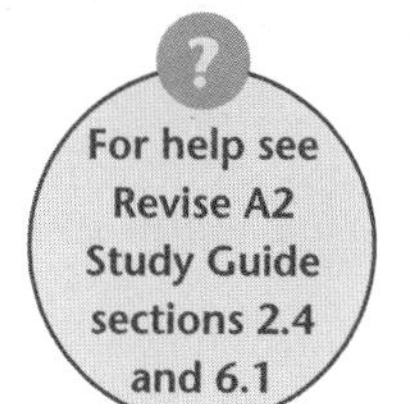

(a) Explain what is meant by **diffraction**. [1]

Diffraction is the spreading of a wave at a gap. ✔

(b) Calculate the wavelength of the microwaves.
Data: $c = 3.0 \times 10^8\ \text{m s}^{-1}$ [2]

$c = f\lambda$

$\lambda = \dfrac{3.0 \times 10^8}{12 \times 10^9} = 0.025\ \text{m}$ ✔

wavelength = 2.5 cm ✔

(c) **(i)** On the axes below, draw a sketch graph to show how the intensity of the diffracted microwaves varies with angle. [2]

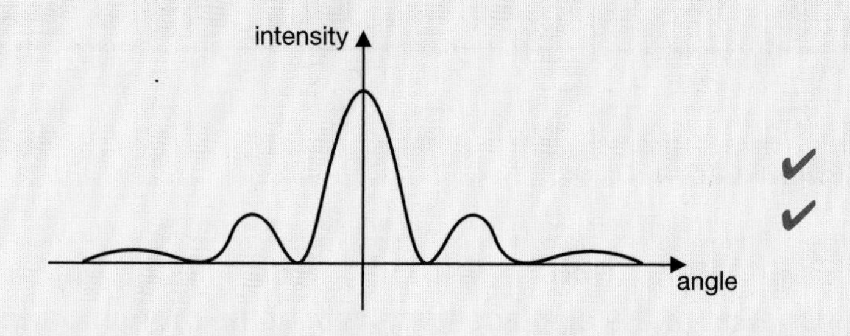

✔
✔

This candidate has correctly sketched the shape of the diffraction pattern and shown that the intensity is zero for specific angles.

(ii) Determine the angle at which the first intensity minimum is observed. [2]

$\theta \approx \dfrac{\lambda}{b} = \dfrac{0.025}{0.15}$ ✔

$\theta \approx 0.167$ radians (9.5°) ✔

Exam practice questions

1 **(a)** Explain what is meant by resonance. Draw a sketch graph of amplitude A of a mechanical oscillator against the forcing frequency f to illustrate your answer. [4]

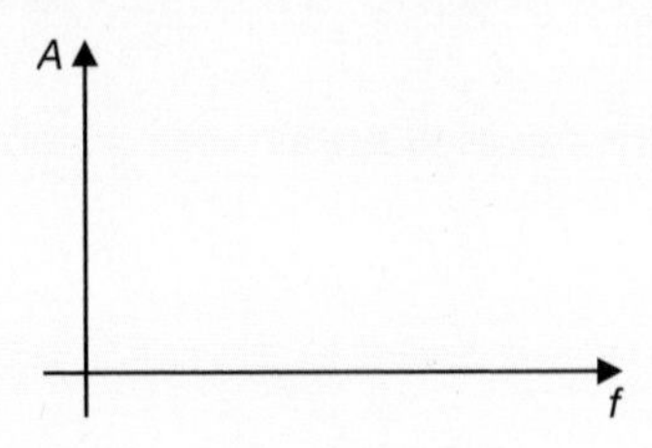

(b) The molecules of HF absorb infra-red radiation very strongly at a wavelength of 2.4×10^{-6} m. Use this information to calculate the natural frequency f_0 with which the molecules of HF vibrate.
Data: $c = 3.0 \times 10^8 \text{ m s}^{-1}$. [3]

[Total: 7]

---

2 **(a)** Define the period of a mechanical oscillator. [1]

(b) Define simple harmonic motion (s.h.m.). [2]

(c) The graph below shows the variation of the displacement x with time t for an object executing s.h.m. The mass of the object is 3.0 kg.

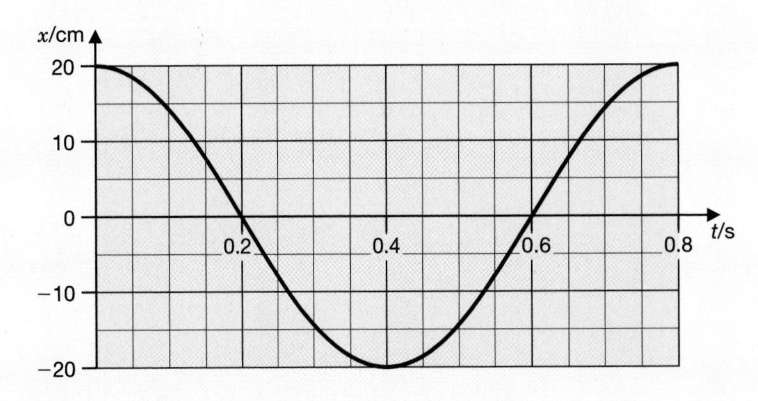

(i) What is the amplitude of the motion? [1]

(ii) On the graph, indicate with a cross (**✗**), a time at which the object has maximum speed. [1]

(iii) Calculate the maximum speed of the object. [3]

(iv) Calculate the maximum force acting on the object. [3]

(v) State one way in which the shape of the graph would change if there was a small amount of damping. [1]

[Total: 12]

Answers on pages 27–31 Answers on pages 27–31 Answers on pages 27–31

3 The diagram shows a small rock suspended from the end of a helical spring which is attached to a mechanical vibrator.

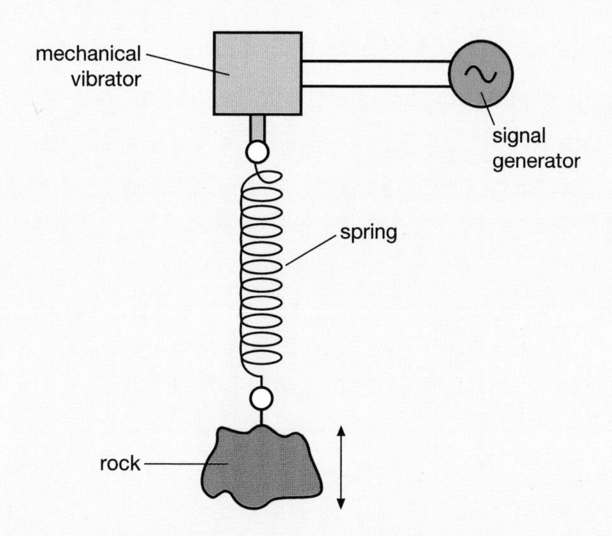

The sketch below shows how the amplitude A of the rock varies with the forcing frequency f of the mechanical vibrator.

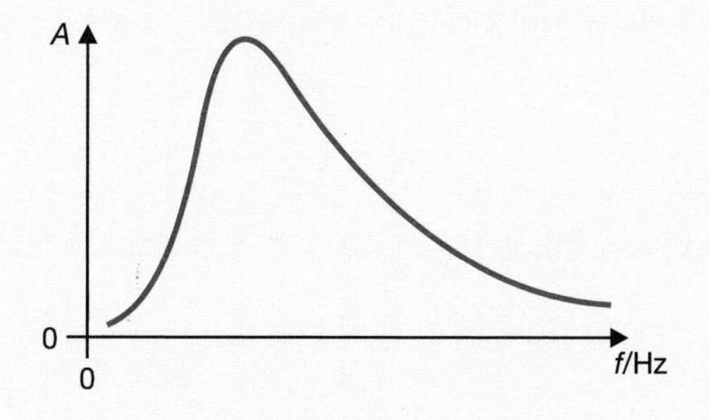

(a) Explain the significance of the peak in the graph above. [1]

(b) The mechanical vibrator is switched off. The rock attached to the spring is allowed to oscillate freely in the vertical plane. The rock undergoes 10 complete oscillations in a time interval of 14.0 s. The spring supporting the rock has a spring constant (force constant) of 8.6 N m^{-1}. Calculate the mass of the rock. [3]

(c) The rock is now replaced by one having half the mass calculated in **(b)**. State and explain the change to the graph of A against f. [2]

[Total: 6]

4 At room temperature, each atom within a particular metal vibrates with a simple harmonic motion. The mass of each atom is 1.1×10^{-25} kg and it may be considered to have an amplitude of 1.3×10^{-11} m and a natural frequency of 7.8×10^{12} Hz.

(a) Calculate the maximum kinetic energy of the atom. [3]

(b) Calculate the total maximum kinetic energy of the atoms in a metal sample of mass 1.0 kg. [3]

(c) On the axes below, sketch a graph to show how the kinetic energy E_k of an atom executing simple harmonic motion varies with time t. [3]

[Total: 9]

5 Light of wavelength 590 nm falls at right angles on a diffraction grating which has 100 lines per mm.

(a) Calculate the separation between the neighbouring lines in metres. [1]

(b) Calculate the angle between the straight through image and the second order image. [3]

(c) Calculate the highest order image that may be observed with this grating and the wavelength of 590 nm. [2]

(d) Suggest **one** reason why a diffraction grating arrangement is more suitable for determining the wavelength of visible light than a two-slit arrangement. [1]

[Total: 7]

6 The diagram shows an arrangement used to calculate the speed of hydrogen ions.

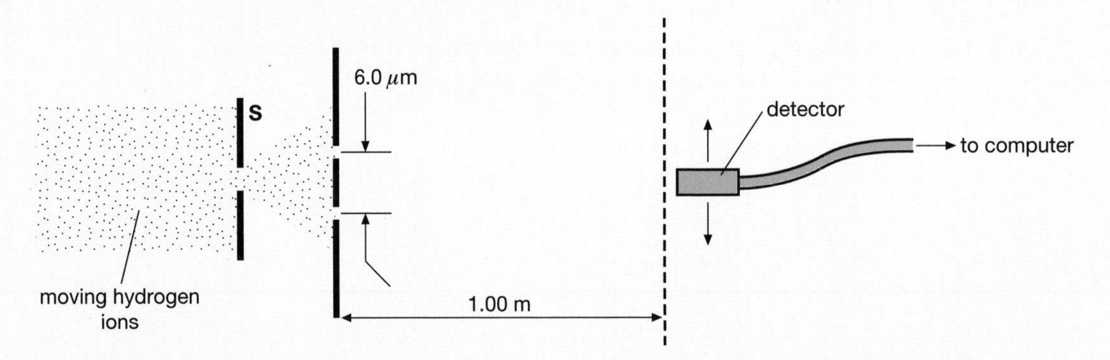

The hydrogen ions emerge from a narrow slit **S** and pass through the double slits to produce an **interference** pattern. The graph on page 25 shows the variation of the number N of ions detected against the distance d measured along the direction in which the detector is moved.

Answers on pages 27–31 **Answers** on pages 27–31 **Answers** on pages 27–31

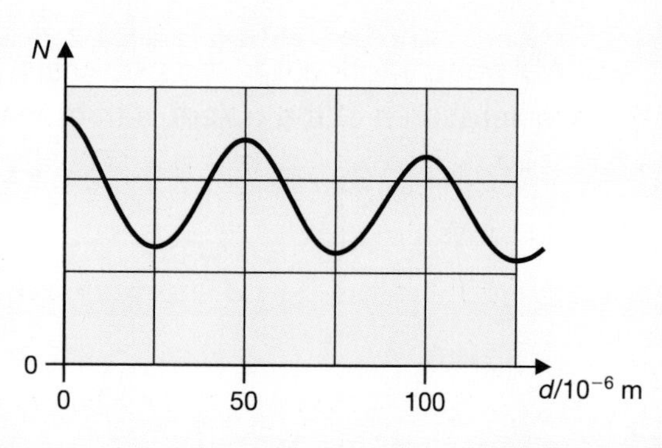

The separation between the double slits is 6.0 μm and the detector is 1.00 m from the double slits.

(a) What does this experiment tell us about moving hydrogen ions? [1]

(b) Use the information from the graph of N against d to calculate the wavelength of a moving hydrogen ion. [3]

(c) Use your answer to **(b)** to determine the speed of a hydrogen ion.
Data: mass of hydrogen ion $= 1.7 \times 10^{-27}$ kg
$h = 6.6 \times 10^{-34}$ J s [3]

[Total: 7]

---

7 A particle executes simple harmonic motion. The motion has an angular frequency of 4.8×10^{4} rad s^{-1}.

(a) Calculate the period of the motion. [2]

(b) On the axes below, show the variation of the displacement x and the velocity v of the particle during three cycles of oscillation. [3]

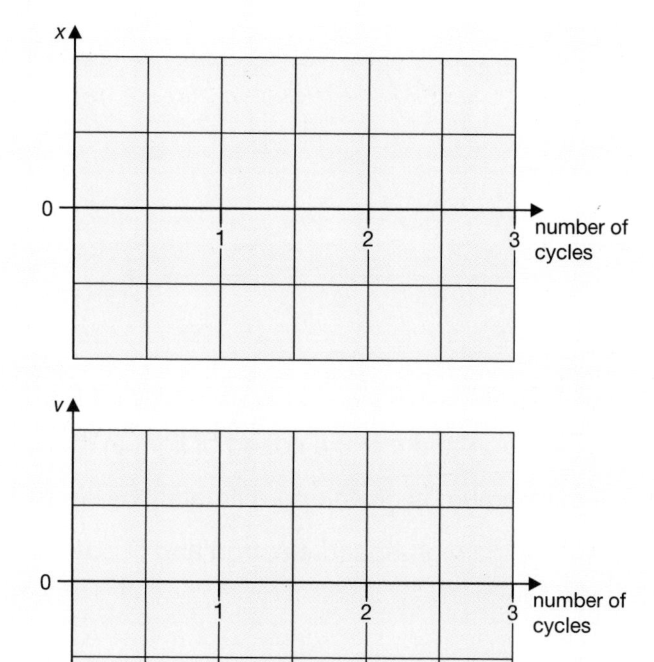

[Total: 5]

Waves and oscillations

8 **(a)** An object executes simple harmonic motion. On the axes below, draw a graph to show the variation of the acceleration a of the object with displacement x. [2]

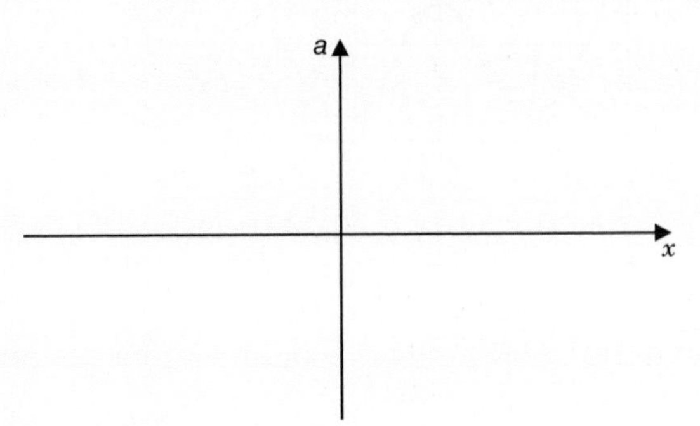

(b) A pendulum bob of mass 65 g is moved a distance of 8.0 cm and then released. The pendulum bob executes simple harmonic motion with a period of 1.6 s. Calculate for the pendulum bob

(i) its maximum speed, [3]

(ii) its displacement 0.50 s after release, [3]

(iii) its maximum gravitational potential energy. [3]

[Total: 11]

---

9 **(a)** Electrons have a **dual** nature. Explain what is meant by this statement. [2]

(b) High-speed electrons are scattered by atomic nuclei and produce a diffraction pattern that resembles the pattern produced by a wave passing through a circular hole. In one experiment, a beam of 280 MeV electrons are scattered by vanadium nuclei. The distribution of the number of electrons N scattered at an angle θ is shown in the sketch below.

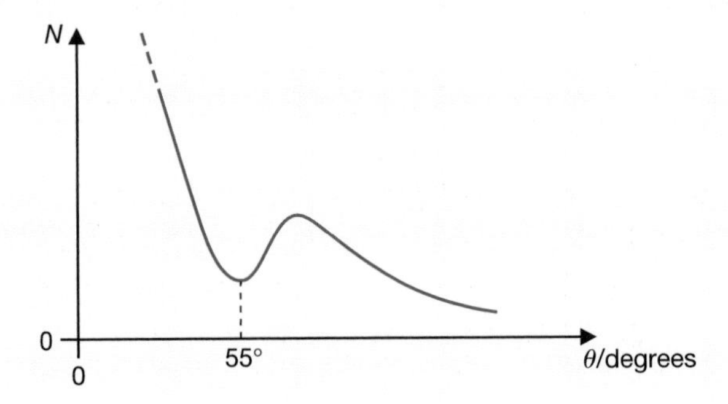

The momentum p of the electron is given approximately by $p \approx \frac{E}{c}$, where E is the kinetic energy of the electron and c is the speed of light in a vacuum. The angle θ for the first diffraction minimum is given approximately by $\sin\theta \approx \frac{\lambda}{D}$, where λ is the de Broglie wavelength of the high-speed electron and D is the diameter of a nucleus.

Data: $h = 6.6 \times 10^{-34}$ J s
$c = 3.0 \times 10^{8}$ m s^{-1}
1 eV $= 1.6 \times 10^{-19}$ J

(i) Calculate the momentum of a 280 MeV electron. [2]

(ii) Calculate the approximate diameter of the vanadium nucleus. [3]

[Total: 7]

Answers on pages 27–31 Answers on pages 27–31 Answers on pages 27–31

Answers

(1) (a) At resonance, the oscillator has maximum amplitude and the forcing frequency is equal to the natural frequency of the oscillator.
The oscillator absorbs maximum energy at resonance.

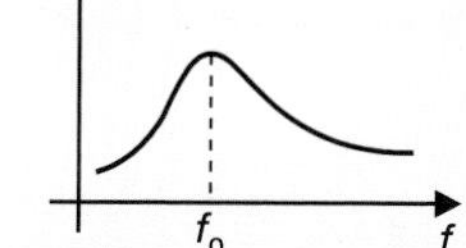

Correct shape of the A against f sketch.

(b) At resonance, the frequency of the infra-red radiation is equal to the natural frequency of HF molecules.

$f_0 = \frac{c}{\lambda}$ $\qquad f_0 = \frac{3.0 \times 10^8}{2.4 \times 10^{-6}}$ $\qquad f_0 = 1.25 \times 10^{14} \approx 1.3 \times 10^{14}$ Hz

examiner's tip

The oscillating electric field of the infra-red radiation is responsible for forcing the HF molecules to vibrate. At resonance, the HF molecules absorb the maximum amount of energy from the incident infra-red radiation.

---

(2) (a) The period is the time taken for one complete oscillation.

(b) The acceleration of the object is directly proportional to its displacement from some fixed point.
The acceleration is directed towards the fixed point.
Therefore, $a \propto -x$.

(c) (i) Amplitude = 20 cm.

examiner's tip

The most common mistake made by candidates is to give the peak-to-peak distance. It is important to be aware that amplitude is equal to the maximum displacement from the equilibrium position of the oscillator.

(ii) A cross (✗) at either $t = 0.2$ s or $t = 0.6$ s.

examiner's tip

The gradient from a displacement–time graph is equal to the velocity of the oscillator. The gradient is maximum at the times indicated above. At 0.2 s the object is moving one way and after one half of an oscillation, it is travelling in the opposite direction with the same speed.

(iii) $T = 0.8$ s

$\omega = \frac{2\pi}{T} = \frac{2\pi}{0.8}$

$\omega = 7.854$ rad s^{-1}

$v_{max} = \omega A = 7.854 \times 0.2$

$v_{max} = 1.57 \approx 1.6$ m s^{-1}

(iv) $a_{max} = \omega^2 A$

$a_{max} = 7.854^2 \times 0.2$

$a_{max} = 12.34$ m s^{-2}

$F = ma = 3.0 \times 12.34$

$F = 37.0 \approx 37$ N

(v) The amplitude of the object will decrease.
For a small amount of friction, the amplitude will decay exponentially with respect to time.

(3) (a) The peak occurs at the resonant frequency. At resonance, the amplitude of oscillation of the rock is a maximum. The natural frequency of oscillation of the rock suspended from the spring is equal to the frequency of the mechanical oscillator.

(b) period, $T = \frac{14.0}{10} = 1.40\text{ s}$

$$T = 2\pi\sqrt{\frac{m}{k}}$$

Therefore $m = \frac{T^2k}{4\pi^2} = \frac{1.40^2 \times 8.6}{4\pi^2} = 0.430\text{ kg}$ (430 g)

(c) the natural frequency $f = \frac{1}{T}$

$$f = \frac{1}{2\pi}\sqrt{\frac{k}{m}}$$

$$f \propto \frac{1}{\sqrt{m}}$$

Hence if the mass is halved, the resonant frequency will increase by a factor $\sqrt{2}$. The graph of A against f will have the same shape, except the peak will shift to a higher frequency.

examiner's tip

You can always calculate the new resonant frequency of the spring–rock system. However, the above solution is quicker and shows understanding of the equation

$$T = 2\pi\sqrt{\frac{m}{k}}.$$

---

(4) (a) $v = \pm\omega\sqrt{A^2 - x^2}$

$v_{max} = \omega A = 2\pi fA$

$v_{max} = 2\pi \times 7.8 \times 10^{12} \times 1.3 \times 10^{-11}$

$v_{max} = 637.2\text{ m s}^{-1} \approx 640\text{ m s}^{-1}$

$KE_{max} = \frac{1}{2}mv^2_{max} = \frac{1}{2} \times 1.1 \times 10^{-25} \times 637.2^2 \approx 2.23 \times 10^{-20}\text{ J}$

(b) number of atoms in 1.0 kg $= \frac{1.0}{1.1 \times 10^{-25}} = 9.091 \times 10^{24} \approx 9.1 \times 10^{24}$

total kinetic energy $= 9.091 \times 10^{24} \times 2.23 \times 10^{-20}$

$= 2.0 \times 10^5\text{ J}$

(c)

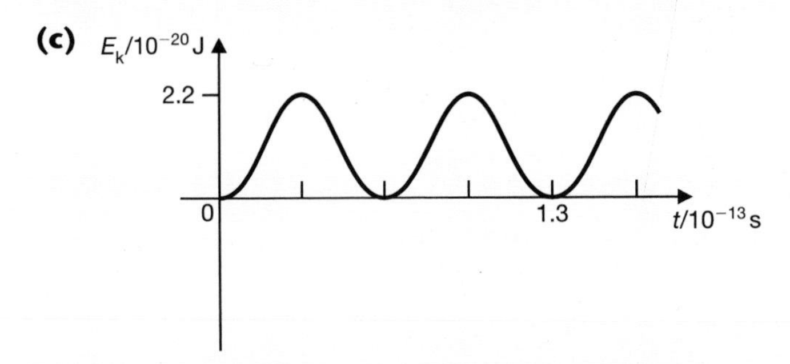

examiner's tip

Since kinetic energy ∝ speed², the graph of E_k against t will always have positive values. For a sketch graph, you must also include some numerical values. The period T of the atom is given by

$$T = \frac{1}{f} = \frac{1}{7.8 \times 10^{12}} \approx 1.3 \times 10^{-13}\text{ s}$$

(5) (a) separation, $d = \frac{0.001}{100}$

$d = 1.0 \times 10^{-5}$ m

(b) $d \sin\theta = n\lambda \quad (n = 2)$

$$\sin\theta = \frac{2 \times 590 \times 10^{-9}}{1.0 \times 10^{-5}} = 0.1180$$

$\theta = 6.8°$

examiner's tip

A common mistake made by candidates in the exams is to use the wavelength of 590 nm and separation d of 0.01 mm in the grating equation $d \sin\theta = n\lambda$. You must convert all quantities into SI units.

(c) $d \sin\theta = n\lambda$

For highest order, θ is 90°. Therefore $\sin\theta = 1$.

$$n_{max} = \frac{d \sin 90°}{\lambda} = \frac{1.0 \times 10^{-5}}{590 \times 10^{-9}} = 16.95$$

The highest order image will be $n = 16$.

examiner's tip

The answer is 16 and not 17. You cannot 'round up' in this question. The 17th order is not seen, hence the highest order must be 16.

(d) A diffraction grating produces images that are sharper than those produced by the two-slit arrangement. The angle θ can therefore be measured more accurately.

---

(6) (a) Hydrogen ions travel through space as 'waves'.

examiner's tip

The double slit arrangement may be used to show interference effects for any wave. Since moving hydrogen ions have a wave-like behaviour, they too, can produce an interference pattern. The wavelength of the ions is given by the de Broglie equation.

(b) 'fringe' separation, $x = 50 \times 10^{-6}$ m

$$\lambda = \frac{ax}{D} = \frac{6.0 \times 10^{-6} \times 50 \times 10^{-6}}{1.00}$$

$\lambda = 3.0 \times 10^{-10}$ m

(c) $\lambda = \frac{h}{mv}$

$$v = \frac{h}{m\lambda} = \frac{6.6 \times 10^{-34}}{1.7 \times 10^{-27} \times 3.0 \times 10^{-10}} \approx 1.3 \text{ km s}^{-1}$$

(7) (a) $\omega = \dfrac{2\pi}{T}$

$$T = \frac{2\pi}{\omega} = \frac{2\pi}{4.8 \times 10^4} = 1.309 \times 10^{-4}\,\text{s}$$

period $\approx 1.3 \times 10^{-4}$ s

(b)

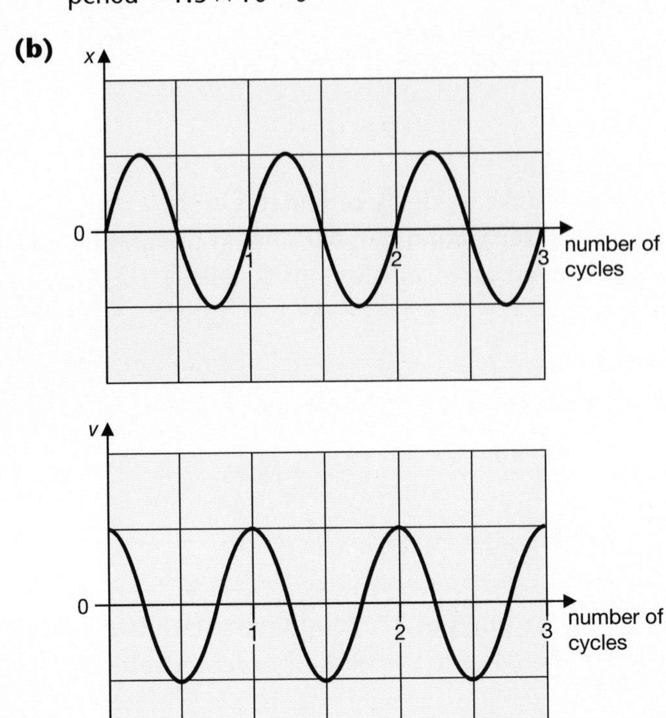

examiner's tip

In this question you may choose any sinusoidal curve for the x against t graph. Once you have done this, the velocity against time graph must be directly related to your displacement against time graph. Remember that the **gradient** from the displacement against time graph is equal to velocity.

---

(8) (a)

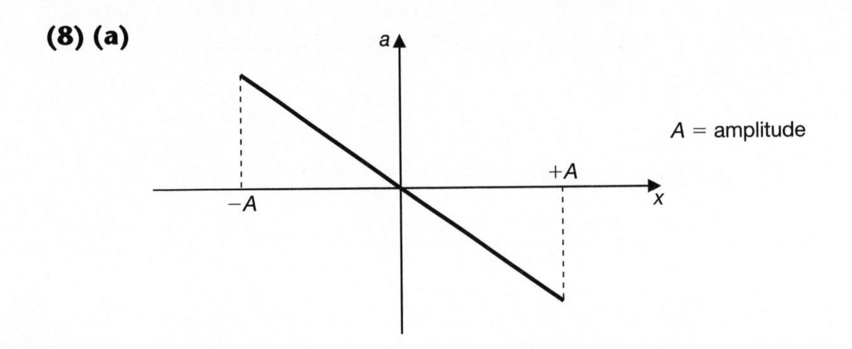

examiner's tip

For simple harmonic motion, the acceleration a is directly proportional to the displacement x and is always directed towards the equilibrium position. That is:

$$a = -\omega^2 x. \qquad (\omega \text{ is a constant})$$

The graph is therefore a straight line graph through the origin and will have a gradient of $-\omega^2$.

(b) (i) $v = \pm\omega\sqrt{A^2 - x^2}$

$$v_{max} = \omega A = (\frac{2\pi}{T})A$$

$$v_{max} = \frac{2\pi \times 0.08}{1.6} = 0.3142\ \text{m s}^{-1} \approx 0.31\ \text{m s}^{-1}$$

(ii) $x = A \cos \omega t$

$$\omega = \frac{2\pi}{T} = \frac{2\pi}{1.6} = 3.928\ \text{rad s}^{-1}$$

$x = 0.08 \times \cos(3.928 \times 0.5) = -0.031$ m (−3.1 cm)

examiner's tip

The correct equation is $x = A \cos \omega t$ and not $x = A \sin \omega t$ because the pendulum bob has a displacement equal to 8.0 cm when the time t is zero. In the calculation, it is very important that you change the 'mode' of the calculator into radians. In the equation $x = A \cos \omega t$, the 'angle' must be radians. Failure to do this will give you a wrong answer.

(iii) $GPE_{max} = KE_{max}$

Therefore

$GPE_{max} = \frac{1}{2} mv^2 = \frac{1}{2} \times 0.065 \times 0.3142^2 \approx 3.2$ mJ

---

(9) (a) Electrons **travel** through space as a **wave** and they **interact** as **particles**.

(b) (i) $p = \frac{E}{c} = \frac{280 \times 10^6 \times 1.6 \times 10^{-19}}{3.0 \times 10^8} = 1.493 \times 10^{-19}\ \text{kg m s}^{-1}$

$p \approx 1.5 \times 10^{-19}\ \text{kg m s}^{-1}$

(ii) $\lambda = \frac{h}{p}$ de Broglie equation

Therefore

$$\lambda \approx \frac{h}{(E/c)} = \frac{hc}{E} = \frac{6.6 \times 10^{-34} \times 3.0 \times 10^8}{280 \times 10^6 \times 1.6 \times 10^{-19}}$$

$\lambda \approx 4.42 \times 10^{-15}$ m

For the first diffraction minimum $\Rightarrow \sin\theta \approx \frac{\lambda}{D}$

$$\sin 55 \approx \frac{4.42 \times 10^{-15}}{D}$$

$$D \approx \frac{4.42 \times 10^{-15}}{\sin 55} \approx 5.4 \times 10^{-15}\ \text{m}$$

Questions with model answers

C grade candidate – mark scored 6/10

1 Define gravitational field strength, g at a point in space. **[1]**

g is the force experienced per unit mass at a point in a gravitational field. ✔

> For help see Revise A2 Study Guide sections 1.4, 3.1 and 3.5

2 The mass of the Earth is 6.0×10^{24} kg and its equatorial radius is 6.4×10^{6} m.

(a) Calculate the surface gravitational field at the equator.
Data: $G = 6.67 \times 10^{-11}$ Nm^2 kg^{-2} **[3]**

$g = -\frac{GM}{r^2}$ ✔

$g = \frac{6.67 \times 10^{-11} \times 6.0 \times 10^{24}}{(6.4 \times 10^6)^2}$ ✔

$g = 9.77$ m s^{-2} (or N kg^{-1}) ✔

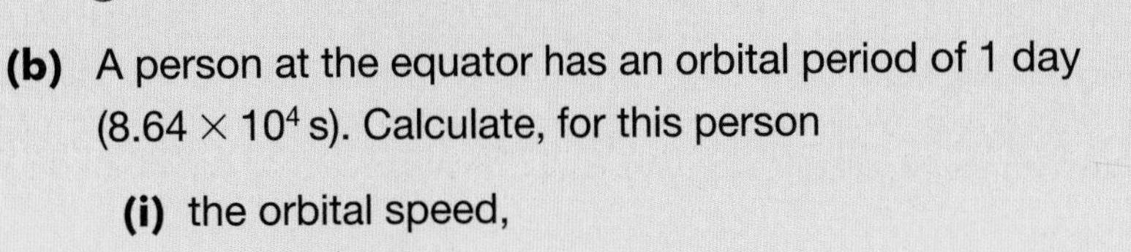

(b) A person at the equator has an orbital period of 1 day (8.64×10^4 s). Calculate, for this person **[2]**

(i) the orbital speed,

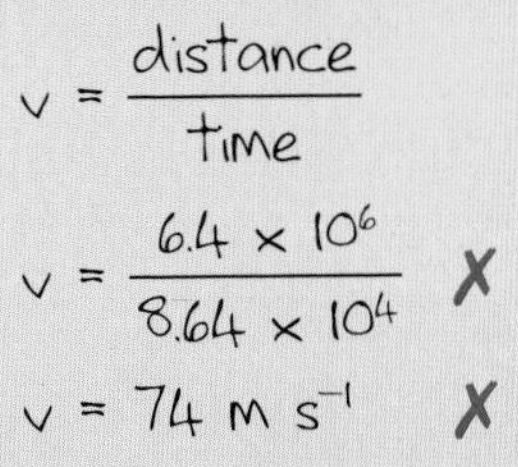

$v = \frac{\text{distance}}{\text{time}}$

$v = \frac{6.4 \times 10^6}{8.64 \times 10^4}$ ✗

$v = 74$ m s^{-1} ✗

(ii) the centripetal acceleration. **[2]**

$a = \omega^2 r$

$a = \left(\frac{2\pi}{8.64 \times 10^4}\right)^2 \times 6.4 \times 10^6$ ✔

$a = 3.4 \times 10^{-2}$ m s^{-1} ✔

(c) Calculate the force exerted by the ground on a 70 kg person at the equator. **[2]**

force = weight = 700 N ✗

Examiner's Commentary

The orbital speed for an object moving in a circle is not equal to the radius of the orbit divided by the period. The speed v is given by

$v = \frac{\text{circumference}}{\text{time}}$

$v = \frac{2\pi r}{T}$ or $v = \omega r$

Correct substitution into the above equation gives an orbital speed of about 470 m s^{-1}

This is designed to be a discriminating question. The candidate's answer is wrong and does not make use of the earlier answers. At the equator, the force R exerted by the ground is given by

$mg - R = ma$

(Newton's second law)

$R = 70 \times (9.77 - 0.034)$

$R \approx 681$ N

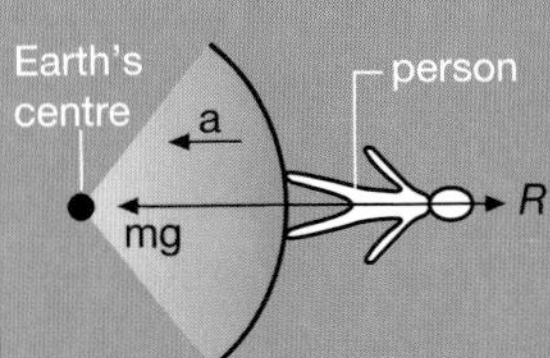

The spinning of the Earth means that the reaction force R provided by the ground is reduced by an amount 'ma'.

GRADE BOOSTER

You cannot afford to have gaps in your knowledge. An extra two marks would have earned the candidate an A grade.

A grade candidate – mark scored 8/9

Examiner's Commentary

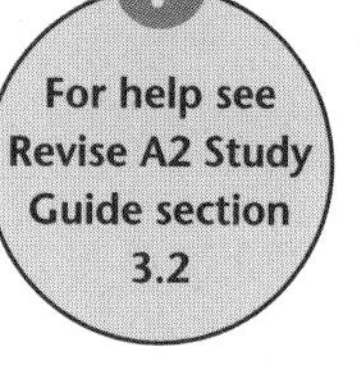

1 The diagram below shows an isolated positively charged metal sphere.

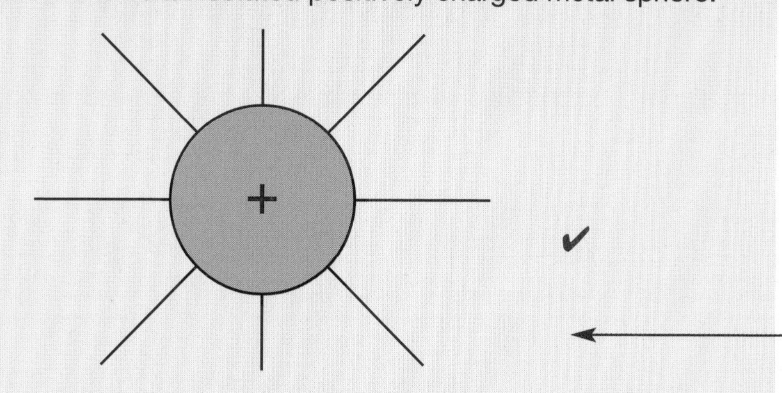

✔

The candidate has lost one 'detail' mark. The radial field pattern is correct, but there is no indication of the direction of the field. The direction of the field is away from the positively charged sphere.

Complete the diagram to show the electric field pattern around the charged sphere. [2]

2 **(a)** Define electric potential at a point in space. [2]

Electric potential is equal to the work done per unit ✔ charge for a charge brought from infinity to that point. ✔

(b) A metal sphere of radius 12 cm is suspended from the ceiling by means of a nylon thread. At a particular moment in time, the surface charge on the sphere is 3.2×10^{-8} C.

(i) Calculate the surface potential for the metal sphere.
Data: $\epsilon_0 = 8.85 \times 10^{-12}$ F m^{-1} [3]

$$V = \frac{Q}{4\pi\epsilon_0 r}$$ ✔

$$V = \frac{3.2 \times 10^{-8}}{4\pi\epsilon_0 \times 1.2 \times 10^{-2}}$$ ✔

V = 2.4 kV ✔

(ii) Explain why the potential on the surface of the sphere decays rapidly during a humid day. [2]

The charge on the sphere decreases. ✔
Since charge ∝ potential, the potential starts to decrease as the charge is 'lost'.
The charge is most likely to be lost through the surface moisture on the thread. ✔

The last answer is superb. The A grade candidate has provided a clear answer with the right blend of mathematics and physics.

Fields

Exam practice questions

1 **(a)** The diagram shows a positively charged sphere placed close to an earthed plate. Draw the electric field pattern between the plate and the charged sphere. [2]

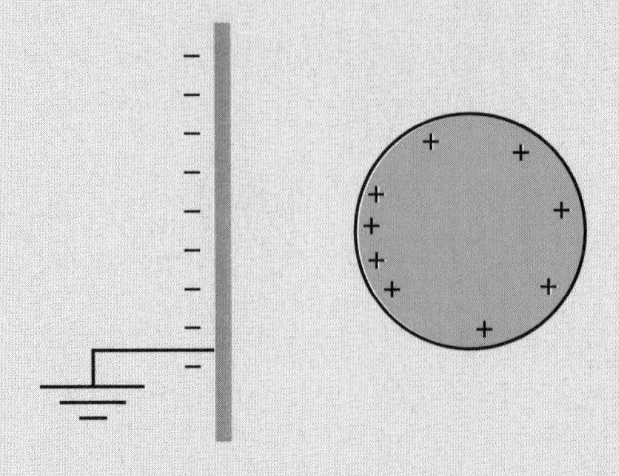

(b) The diagram shows a pair of parallel metal plates separated by a distance of 1.0 cm. The potential difference across the plates is 3.0 kV.

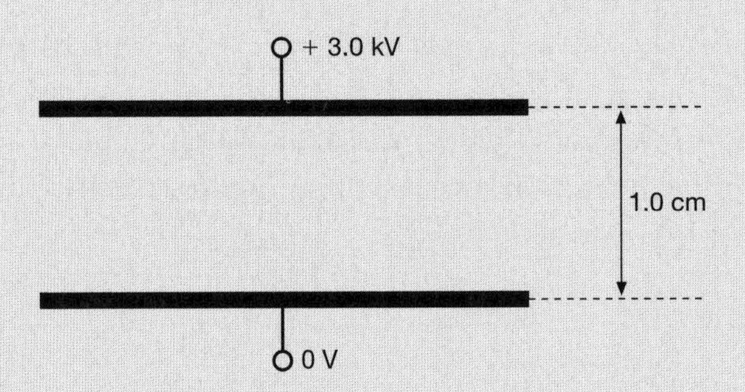

(i) On the diagram, draw a line representing the 1.0 kV **equipotential**. [2]

(ii) The electric field between the plates may be considered to be uniform. Calculate the field strength E between the plates. [3]

(iii) Calculate the force experienced by an electron between the plates due to the electric field of the plates.
Data: $e = 1.6 \times 10^{-19}$ C [2]

[Total: 9]

---

2 Outline the similarities and the differences between electric and gravitational fields for **point** objects. [4]

[Total: 4]

Answers on pages 40–45 Answers on pages 40–45 Answers on pages 40–45

3 The diagram shows a 100 μF capacitor connected in a circuit.

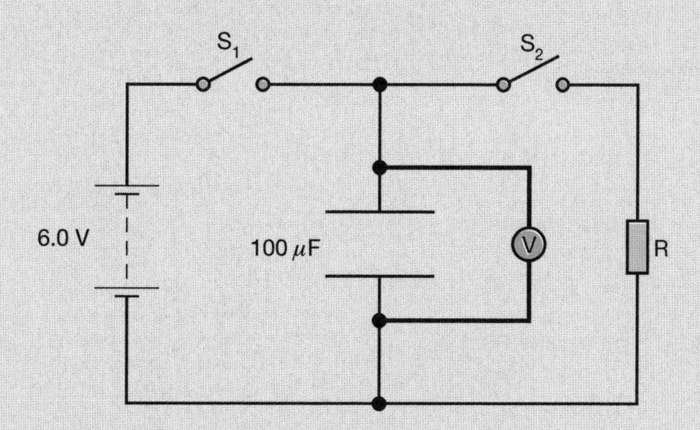

(a) With switch S_1 closed and S_2 open, state what happens to the potential difference (p.d.) across the capacitor. The capacitor is initially uncharged. [1]

(b) The switch S_1 is closed for a short period of time and then opened again. The graph below shows the variation of the p.d., *V* across the capacitor with time *t* when the switch S_2 is closed.

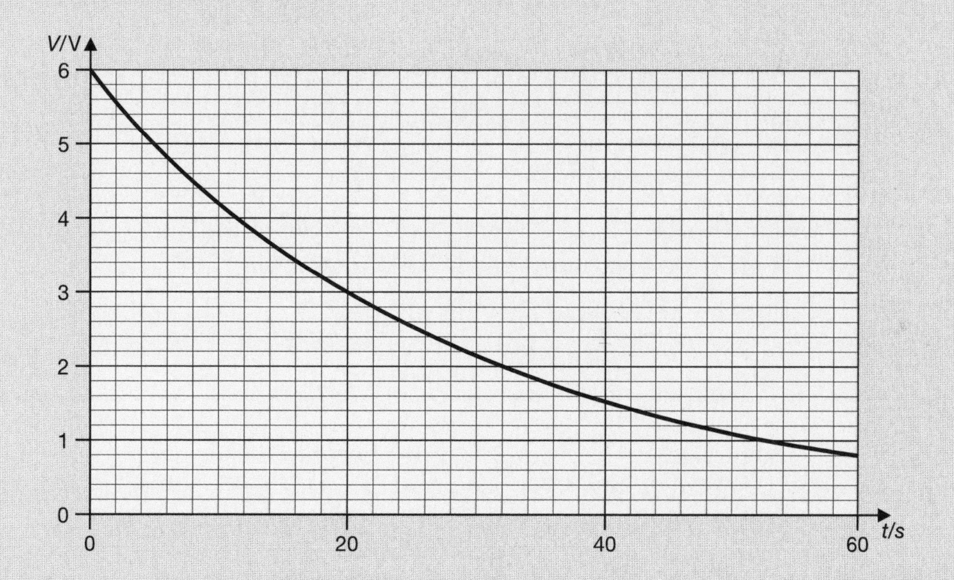

(i) Calculate the energy stored by the fully charged capacitor. [3]

(ii) Determine the time constant of the circuit. [1]

(iii) Use your answer to **(b)(ii)** to calculate the resistance *R* of the resistor. [3]

(iv) The p.d. across the capacitor decays exponentially. What does **exponential decay** mean? [1]

[Total: 9]

Fields

4 The diagram below shows a capacitor designed from two identical sheets of aluminium foil and some paper.

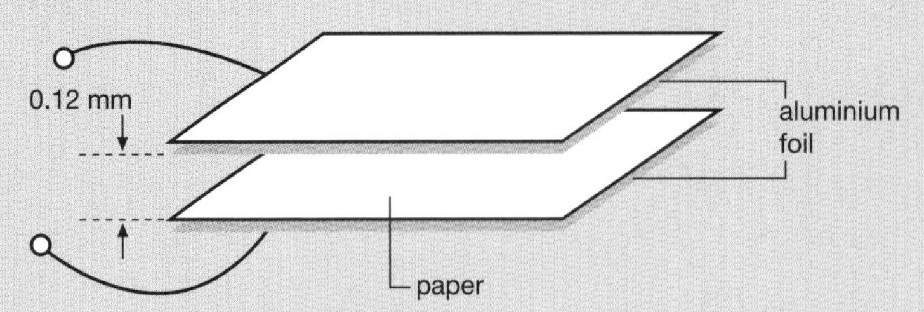

Each aluminium foil has a surface area of $8.0 \times 10^{-3}\ \text{m}^2$ and the thickness of the paper is 0.12 mm.

(a) Show that the capacitance *C* of the capacitor is equal to 1.4 nF.
Data: $\epsilon_0 = 8.85 \times 10^{-12}\ \text{F m}^{-1}$
relative permittivity of paper = 2.4 **[2]**

(b) The 1.4 nF capacitor is charged to a potential difference of 60 V.

(i) Calculate the magnitude of the charge *Q* on one of the capacitor plates. **[2]**

(ii) The charged capacitor is then connected across another identical 1.4 nF uncharged capacitor via a switch **S** as shown in the diagram.

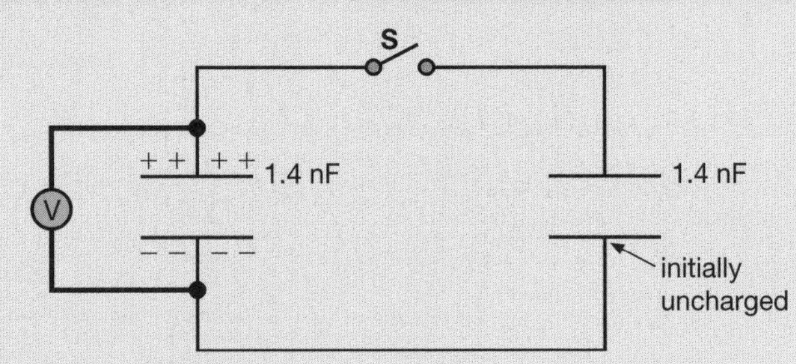

The switch **S** is closed.

1. Explain why the voltmeter reading decreases. **[2]**

2. Calculate the new reading on the voltmeter. **[3]**

[Total: 9]

---

5 The diagram below shows the initial path of an electron observed in a nuclear particle detector. The electron has been created along with another particle, not shown here, in the detector at point **A**. There is a uniform magnetic field perpendicular to the plane of the diagram.

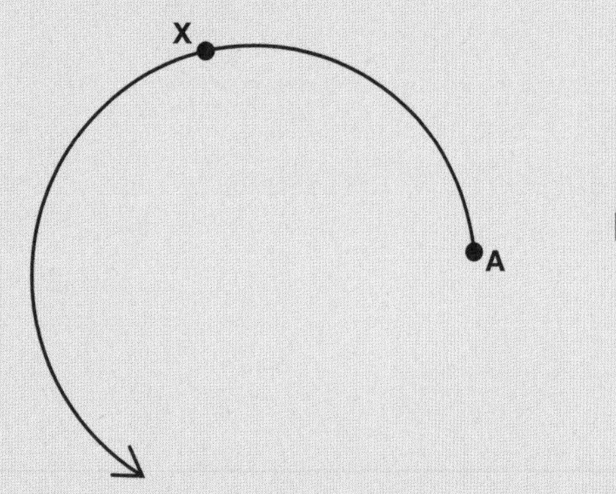

Data: mass of electron = 9.1×10^{-31} kg
$e = 1.6 \times 10^{-19}$ C

(a) The speed of the electron at **X** is $1.0 \times 10^8\ \text{m s}^{-1}$. The radius of curvature of the electron path is 0.040 m. Calculate the magnitude of the force on the electron. **[3]**

(b) Calculate the magnitude of the magnetic flux density *B* in the detector. Give a suitable unit for your answer. **[4]**

[OCR – Syllabus A June 2003]

[Total: 7]

6 **(a)** The planet Jupiter may be considered to be a uniform isolated sphere of radius 7.1×10^7 m. Its mass may be assumed to be concentrated at the centre. At its surface, the gravitational field strength is $25\ \text{N kg}^{-1}$.
Data: $G = 6.67 \times 10^{-11}\ \text{N m}^2\ \text{kg}^{-2}$

Calculate

(i) its mass, [3]

(ii) its mean density. [3]

(b) A satellite is placed in geostationary orbit around the Earth.
Data: mass of Earth = 6.0×10^{24} kg
$G = 6.67 \times 10^{-11}\ \text{N m}^2\ \text{kg}^{-2}$

(i) Explain what is meant by a geostationary orbit. [1]

(ii) Calculate the orbital radius of a geostationary satellite around the Earth. [4]

[Total: 11]

---

7 The moon has mass 7.4×10^{22} kg and mean radius 1.7×10^6 m. It may be considered to be an isolated uniform sphere with its mass concentrated at its centre.
Data: $G = 6.67 \times 10^{-11}\ \text{N m}^2\ \text{kg}^{-2}$

(a) Calculate the gravitational potential at the surface of the Moon. [3]

(b) **(i)** Show that the escape velocity v of any object from the surface of the moon is given by

$$v = \sqrt{\frac{2GM}{R}}$$

where M is the mass of the moon and R is its radius. [3]

(ii) Calculate the escape velocity of an object from the surface of the moon. [2]

[Total: 8]

Fields

---

8 The diagram below shows an arrangement of capacitors.

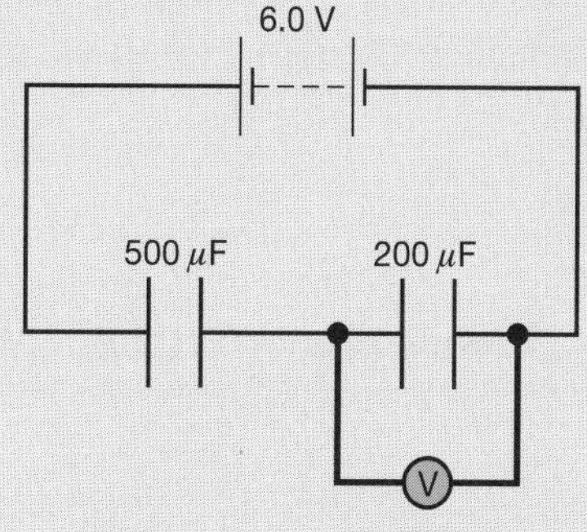

The voltmeter placed across the 200 μF capacitor has an infinite resistance.

(a) Define the capacitance of a capacitor. [1]

(b) Calculate

(i) the total capacitance of the circuit, [3]

(ii) the voltmeter reading. [3]

[Total: 7]

Answers on pages 40–45 Answers on pages 40–45 Answers on pages 40–45

9 **(a)** State Coulomb's law. [1]

(b) Two small metal spheres are suspended from a point by insulated threads. Each of the spheres is given an equal amount of positive charge and they repel each other. Their equilibrium state is shown in the diagram below.

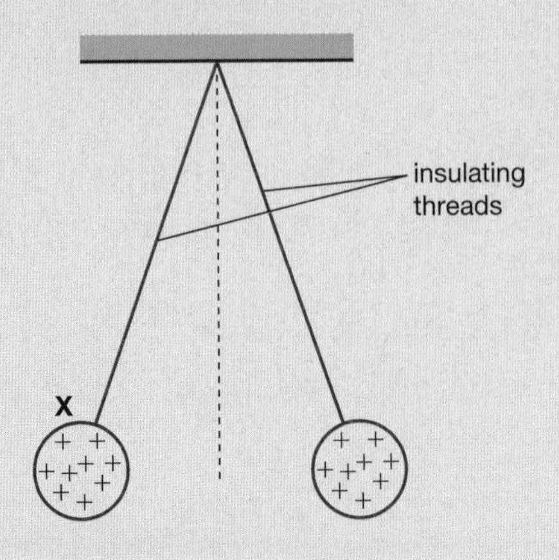

(i) On the diagram above, show all the forces experienced by the metal sphere **X**. [3]

(ii) The electrical force acting on each metal sphere is 8.4×10^{-5} N. In its equilibrium state, the centre-to-centre separation of the two spheres is 32 mm. The radius of each sphere is 12 mm.
Data: $\epsilon_0 = 8.85 \times 10^{-12}\ \text{F m}^{-1}$

Calculate

1. the charge on each metal sphere, [3]
2. the electrical potential at the surface of the sphere. [3]

[Total: 10]

Fields

 Answers on pages 40–45 Answers on pages 40–45 Answers on pages 40–45

10 An electron travelling at a speed of 7.0×10^6 m s^{-1} enters an evacuated region where there is both a uniform magnetic field and a uniform electric field, as shown in the diagram below.

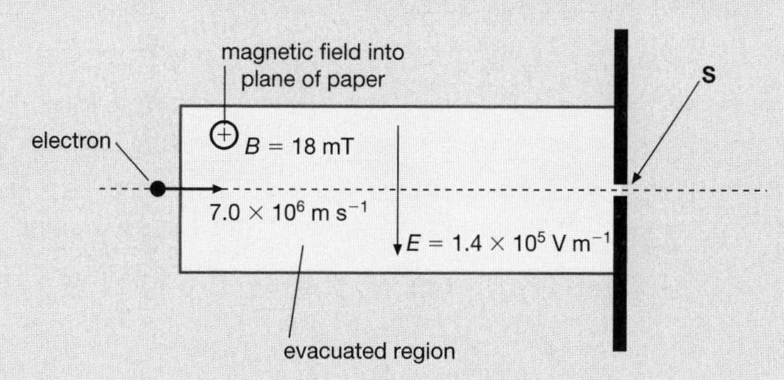

Data: $e = 1.6 \times 10^{-19}$ C

(a) Determine the magnitude and direction of the force experienced by the electron due to

(i) the magnetic field, [2]

(ii) the electric field. [2]

(b) Explain why the electron does not emerge from the slit **S**. [1]

(c) For the magnetic field and electric field given above, calculate the speed of the electron that would emerge from the slit. [3]

[Total: 8]

---

11 **(a)** State Faraday's law of electromagnetic induction. [2]

(b) A coil is placed in a uniform magnetic field. The magnetic field is at right angles to the plane of the coil. For this coil, define

(i) magnetic flux, [1]

(ii) magnetic flux linkage. [1]

(c) A flat circular coil of 500 turns and a mean cross-sectional area 8.0×10^{-4} m^2 is connected to an ammeter. The total resistance of the ammeter and the coil is 0.80 Ω. The plane of the coil is placed at right angles to a uniform magnetic field of field strength 120 mT. The coil is removed from the magnetic field in a time of 30 ms. Calculate the magnitude of the average induced

(i) electromotive force (e.m.f.) across the ends of the coil, [3]

(ii) current in the coil. [3]

[Total: 10]

Answers on pages 40–45 Answers on pages 40–45 Answers on pages 40–45

Answers

(1) (a) Correct field pattern.
Correct field direction shown.

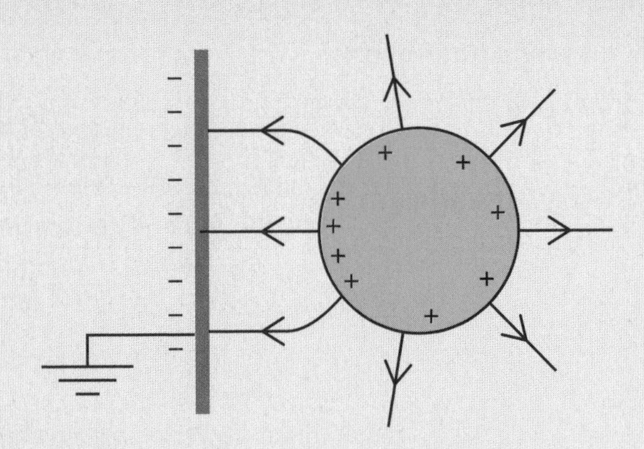

examiner's tip

The radial field pattern of the charged sphere is distorted by the earthed plate. The electric field lines are normal to both the plate and the surface of the sphere.

(b) (i) Straight line between the plate.
This line lies at a distance of $\frac{1}{3}$ cm from the 0 V plate and curves in the region beyond the plates.

examiner's tip

An equipotential is a line (or a surface) of equal potential. Since the field strength between the parallel plates is uniform, the potential *V* must change uniformly in the space between the plates. The 1.0 kV equipotential must therefore lie at a distance of $\frac{1}{3}$ cm from the negative plate. Beyond the parallel plates, the 1.0 kV equipotential is curved.

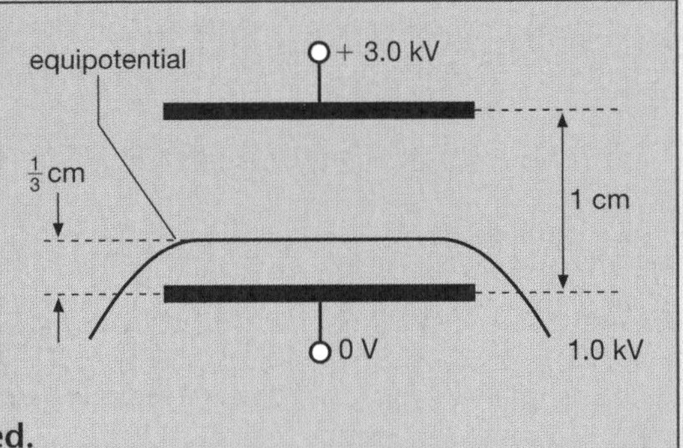

(ii) $E = -$ potential gradient

$$E = \frac{V}{d} \quad \text{(uniform field)}$$

$$E = \frac{3.0 \times 10^3}{0.01}$$

$$E = 3.0 \times 10^5\ \text{V m}^{-1}$$

(iii) $F = Eq$

$F = 3.0 \times 10^5 \times 1.6 \times 10^{-19}$

$F = 4.8 \times 10^{-14}$ N

---

(2) Electric field is to do with **charges**.
Gravitational field is to do with objects having **mass**.
Both fields obey the inverse square law with distance.
Both fields are to do with 'force or action at a distance'.

examiner's tip

There are a lot of other comments that could be added here. The points made above outline the main differences and similarities between the two fields.

(3) **(a)** The p.d. increases to 6.0 V.

(b) **(i)** $E = \frac{1}{2}V^2C$

$E = \frac{1}{2} \times 6.0^2 \times 100 \times 10^{-6}$

$E = 1.8 \times 10^{-3}$ J

(ii) time constant $\approx$ 30 s

examiner's tip

In a time equal to the time constant of the circuit, the p.d. decreases to a value of about 0.368 (which is e^{-1}) of its initial value. This value may be obtained from the graph by reading off the time when the p.d. is equal to

$$0.368 \times 6.0 \approx 2.2\text{ V}$$

(iii) time constant = CR

$CR \approx 30$ s

Therefore $R \approx \dfrac{30}{100 \times 10^{-6}}$

$R \approx 3.0 \times 10^5\ \Omega$

(iv) In a given interval of time, the p.d. across the capacitor decreases by the same factor.

Or

At any time, the **rate** of decrease in the p.d. is directly proportional to the p.d. at that time.

---

(4) (a) $C = \dfrac{\epsilon_0 \epsilon_r A}{d}$

$C = \dfrac{8.85 \times 10^{-12} \times 2.4 \times 8.0 \times 10^{-3}}{0.12 \times 10^{-3}}$

$C = 1.42 \times 10^{-9}$ F

examiner's tip

The area A is the area of overlap between the two aluminium sheets. A common mistake made by candidates is to use the total area of the two capacitor plates.

(b) **(i)** $Q = VC$

$Q = 60 \times 1.42 \times 10^{-9}$

$Q = 8.52 \times 10^{-8} \approx 8.5 \times 10^{-8}$ C

+Q −Q

examiner's tip

The positive plate acquires a charge $+Q$ and the negative plate a charge $-Q$. The total charge on both plates is zero. A capacitor is a device that separates charge. In the question, you are required to calculate the magnitude of the charge on one of the plates. This charge is given by the equation

$$Q = VC$$

(ii) 1. The charge on the original capacitor decreases as it is **shared** with the other capacitor.
Since $V \propto Q$, the p.d. across the capacitor decreases.

2. C_T = total capacitance
$C_T = C_1 + C_2 = 2.84\ \text{nF}$

$V = \frac{Q}{C}$ (applied to the whole circuit)

$$V = \frac{8.52 \times 10^{-8}}{2.84 \times 10^{-9}}$$

$V = 30\ \text{V}$

examiner's tip

The total capacitance of the circuit is doubled. Since

$$V \propto \frac{1}{C}$$

for the same total charge, the p.d. across the capacitors will be halved. Remember that the total charge in the circuit is conserved.

---

(5) (a) force $= \frac{mv^2}{r}$

$$F = \frac{9.1 \times 10^{-31} \times (1.0 \times 10^8)^2}{0.040} = 2.275 \times 10^{-13}\ \text{N}$$

$F \approx 2.3 \times 10^{-13}\ \text{N}$

examiner's tip

In this question you must realise from the onset that the force acting on the electron is a centripetal force because it is moving in a circular path.

(b) $F = BQv$

$$B = \frac{F}{Qv} = \frac{2.275 \times 10^{-13}}{1.6 \times 10^{-19} \times 1.0 \times 10^8} = 0.014$$

unit: tesla (T)

(6) (a) (i) $g = \frac{GM}{r^2}$

$$M = \frac{gr^2}{G} = \frac{25 \times (7.1 \times 10^7)^2}{6.67 \times 10^{-11}} = 1.889 \times 10^{27}\ \text{kg}$$

$M \approx 1.9 \times 10^{27}$ kg

(ii) density $= \frac{\text{mass}}{\text{volume}}$

$$\rho = \frac{1.889 \times 10^{27}}{\frac{4}{3}\pi \times (7.1 \times 10^7)^3} \approx 1300\ \text{kg m}^{-3}$$

(b) (i) The satellite remains above the same point on the Earth's surface. The period of the satellite is equal to the rotational period of the Earth (1 day).

(ii) $F = ma$

$\frac{GMm}{r^2} = ma$ (M = mass of the Earth and m = mass of satellite)

$\frac{GM}{r^2} = a = \frac{v^2}{r}$

$\frac{GM}{r^2} = \frac{(2\pi r/T)^2}{r}$ (T = period of satellite)

$T^2 = (\frac{4\pi^2}{GM})r^3$

$$r = \sqrt[3]{\frac{GMT^2}{4\pi^2}} = \sqrt[3]{\frac{6.67 \times 10^{-11} \times 6.0 \times 10^{24} \times (24 \times 60 \times 60)^2}{4\pi^2}}$$

$r = 4.2 \times 10^7$ m

examiner's tip

This question requires a good knowledge of both gravitational fields and circular motion. It would be worthwhile learning all steps in the proof of 'Kepler's third law': $T^2 = (\frac{4\pi^2}{GM})r^3$

---

(7) (a) potential $= -\frac{GM}{r}$

$$\text{potential} = -\frac{6.67 \times 10^{-11} \times 7.4 \times 10^{22}}{1.7 \times 10^6} \approx -2.9\ \text{MJ kg}^{-1}$$

(b) (i) For an object to escape from the Moon's surface,
kinetic energy = magnitude of potential energy at surface

$\frac{1}{2}mv^2 = \frac{GMm}{r}$

$\frac{v^2}{2} = \frac{GM}{r}$ (The escape velocity is independent of mass m of object.)

Hence, $v = \sqrt{\frac{2GM}{R}}$.

(ii) $v = \sqrt{\frac{2GM}{R}}$

$$v = \sqrt{\frac{2 \times 6.67 \times 10^{-11} \times 7.4 \times 10^{22}}{1.7 \times 10^6}}$$

$v \approx 2.4 \times 10^3\ \text{m s}^{-1}$

(8) (a) $C = \dfrac{Q}{V}$

where C is the capacitance of the capacitor, Q the charge on the capacitor and V the potential difference across the capacitor.

(b) (i) $\dfrac{1}{C_T} = \dfrac{1}{C_1} + \dfrac{1}{C_2}$

$\dfrac{1}{C_T} = \dfrac{1}{200} + \dfrac{1}{500}$

$C_T = \dfrac{200 \times 500}{200 + 500} \approx 143\ \mu F$

examiner's tip

In exams, candidates often write the wrong equation '$C_T = C_1 + C_2$', for the total capacitance of two capacitors in series because of confusion with the equation '$R_T = R_1 + R_2$' for resistors in series.

(ii) total charge, $Q = VC_T$

$Q = 6.0 \times 143 \times 10^{-6} \approx 8.58 \times 10^{-4}\ C$

$V = \dfrac{Q}{C} = \dfrac{8.58 \times 10^{-4}}{200 \times 10^{-6}} \approx 4.3\ V$

examiner's tip

Each capacitor stores the same charge when connected in series. The total charge is the charge stored by each of the capacitors.

---

(9) (a) The force between two point charges is directly proportional to the product of the charges and inversely proportional to the square of their separation.

(b) (i)

tension in thread

X

electrical force

weight

(ii) 1. $F = \dfrac{Qq}{4\pi\epsilon_0 r^2} = \dfrac{Q^2}{4\pi\epsilon_0 r^2} \qquad (Q = q)$

$Q = \sqrt{4\pi\epsilon_0 r^2 F} = \sqrt{4\pi \times 8.85 \times 10^{-12} \times 0.032^2 \times 8.4 \times 10^{-5}}$

$Q = 3.093 \times 10^{-9}\ C \approx 3.1\ nC$

2. $V = \dfrac{Q}{4\pi\epsilon_0 r}$

$V = \dfrac{3.093 \times 10^{-9}}{4\pi \times 8.85 \times 10^{-12} \times 0.012} \approx 2.3 \times 10^3\ V$

examiner's tip

Do not forget that r in this equation is the radius of the sphere and not the separation between the spheres.

(10) (a) (i) $F = BQv$

$F = 0.018 \times 1.6 \times 10^{-19} \times 7.0 \times 10^{6} = 2.016 \times 10^{-14}\,\text{N}$

$F \approx 2.0 \times 10^{-14}\,\text{N}$

Direction of force: Down the page

examiner's tip

The direction of the force is given by Fleming's left hand rule. You must remember that the second finger of the left hand points in the direction of the conventional current, which is opposite to the direction of travel of the negatively charged electron.

(ii) $F = EQ$

$F = 1.4 \times 10^{5} \times 1.6 \times 10^{-19} = 2.24 \times 10^{-14}\,\text{N}$

$F \approx 2.2 \times 10^{-14}\,\text{N}$

Direction of force: Up the page (opposite to the direction of the electric field)

(b) The force due to the electric field is greater than the force due to the magnetic field; hence the electron will not travel in a straight line and it will therefore not emerge from the slit.

(c) force due to electric field = force due to magnetic field

$EQ = BQv$

$E = Bv$

$$v = \frac{E}{B} = \frac{1.4 \times 10^{5}}{0.018} \approx 7.8 \times 10^{6}\,\text{m}\,\text{s}^{-1}$$

---

(11) (a) Faraday's law: The magnitude of the e.m.f. induced in a circuit is directly proportional to the rate of change of magnetic flux linkage in the circuit.

examiner's tip

It is vital to appreciate that the law is concerned with electromotive force (e.m.f.) and not voltage or potential difference. In the definition, you can substitute the word *proportional* with *equal* without losing any marks in the exams.

(b) (i) magnetic flux = magnetic flux density × cross-sectional area of coil

$\phi = BA$ (The field is normal to the plane of the coil.)

(ii) magnetic flux linkage = number of turns × magnetic flux

magnetic flux linkage = NBA

(c) (i) initial magnetic flux linkage = $NBA = 500 \times 0.120 \times 8.0 \times 10^{-4}$

$= 0.048\,\text{Wb}$

final magnetic flux linkage = 0

change in magnetic flux linkage = 0 − 0.048 = −0.048 Wb

e.m.f. = rate of change of magnetic flux linkage

$$\text{e.m.f.} = \frac{\Delta(NAB)}{\Delta t} = \frac{0.048}{0.030} = 1.6\,\text{V} \quad \text{(magnitude only)}$$

(ii) $$\text{current} = \frac{\text{e.m.f.}}{\text{total resistance}}$$

$$\text{current} = \frac{1.6}{0.80} = 2.0\,\text{A}$$

Fields

Chapter 4 Particle physics, cosmology/astrophysics and medical physics

Questions with model answers

C grade candidate – mark scored 6/10

1 Explain the terms in italics below, giving an example in each case.

For help see Revise A2 Study Guide section 4.4

Antiparticle. [2]

These are particles with the same mass but with opposite charge. ✔

Annihilation. [3]

When particles collide and destroy each other. ✗

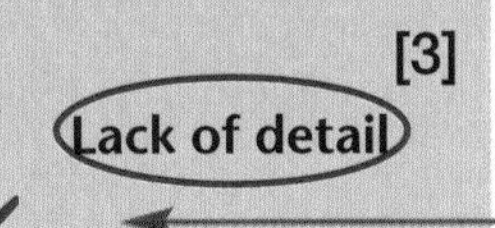

The collision produces electromagnetic radiation. ✔

2 The diagram below illustrates the quark model for the proton.

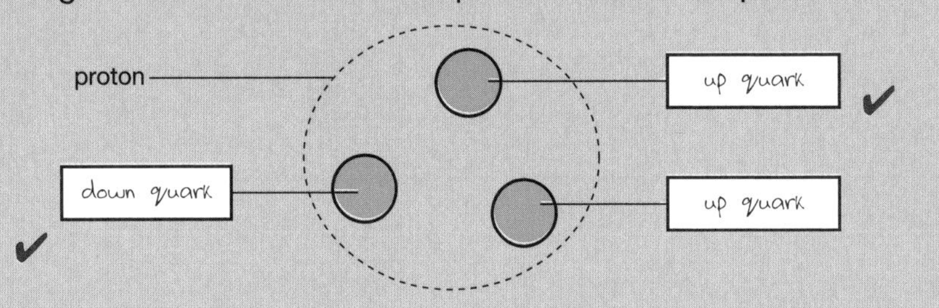

✔ ✔

Complete the diagram by labelling the type of quarks within the proton. [2]

3 (a) The radius r of the nucleus is given by

$$r = r_0 A^{1/3}$$

where r_0 is a constant and A is the nucleon number.

Use this expression to determine the ratio [2]

$$\frac{\text{radius of } ^{235}\text{U nucleus}}{\text{radius of } ^{12}\text{C nucleus}}$$

$$\text{ratio} = \frac{\sqrt[3]{235}}{\sqrt[3]{12}}$$ ✔

$$\text{ratio} = \frac{6.17}{2.29} = 2.7$$ ✔

(b) Suggest why the density of nuclear material is very much greater than the density of ordinary matter, like a piece of rock. [1]

A rock is made from atoms. The radius of the atom is larger than the nucleus. ✗

Examiner's Commentary

The candidate has had success with defining the first term, but there is no named antiparticle. As a result, the candidate has lost one mark. A good example of an antiparticle is the positron, which has the same mass as that of an electron, but with a positive charge of $+1.6 \times 10^{-19}$ C.

A proton consists of two up quarks and a single down quark. The up quark has a charge $+\frac{2}{3}$e and the down quark $-\frac{1}{3}$e, making a total charge of +e for the proton.

The candidate's answer lacks completeness. The candidate has failed to realise that roughly the **same** amount of mass is contained in a much **smaller** volume of the nucleus – the electrons are not massive compared to a proton or a neutron, therefore their contribution to the mass of the atom is negligible. Hence the density of the nucleus is much greater than that of the atoms. Incidentally, the radius of most atoms is ~ 10^{-10} m, whereas the radius of most nuclei is ~ 10^{-15} m.

GRADE BOOSTER

Always use the marks allocated for the part questions as a guide to the numbers of separate steps or comments needed to score full marks.

A grade candidate – mark scored 12/12

Examiner's Commentary

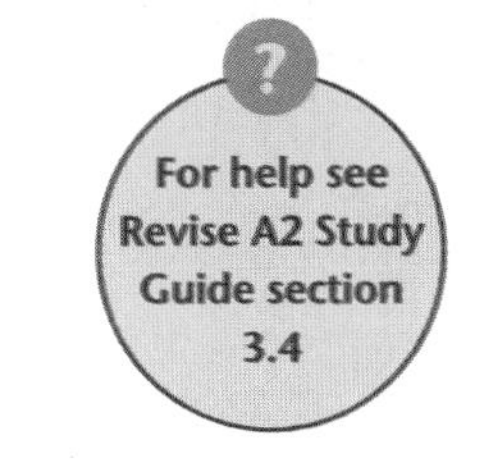

1 Complete the table below by naming the remaining three fundamental interactions (forces) and the exchange particles associated with the interactions. [6]

interaction	exchange particle	
gravitational	graviton	
electromagnetic	photons	✔ ✔
strong nuclear force	gluons	✔ ✔
weak nuclear force	W^+, W^- and Z particles	✔ ✔

The weak nuclear force is mediated through the particles listed above by the candidate. These particles are collectively referred to as the Intermediate Vector Bosons.

2 Distinguish between leptons and hadrons. [4]

Hadrons are made up of quarks. ✔
Hadrons are effected by the strong nuclear force. ✔
Leptons are fundamental particles. ✔
Leptons feel the weak nuclear force. ✔

3 The Feymann diagram below illustrates the decay of a neutron.

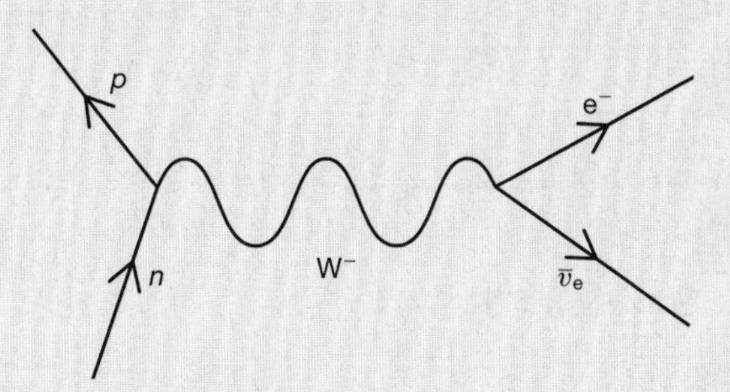

With reference to the diagram, explain each stage of the decay. [2]

A neutron decays into a proton and the W^- boson. ✔
In time, the W^- boson decays into an electron and an anti-neutrino. ✔
The decay shown above is that for β-decay within a nucleus.

Exam practice questions

1 **(a)** **(i)** Underline the particles in the following list that may be affected by the weak interaction.

positron neutron photon neutrino positive pion **[2]**

(ii) Underline the particles in the following list that may be affected by the electromagnetic force.

electron antineutrino proton neutral pion negative muon **[2]**

(b) A positive muon may decay in the following way

$$\mu^+ \Rightarrow e^+ + \nu_e + \bar{\nu}_\mu.$$

(i) Exchange each particle for its corresponding antiparticle and complete the equation to show how a negative muon may decay. **[1]**

$$\mu^- \Rightarrow$$

(ii) Give one difference and one similarity between a negative muon and an electron. **[2]**

[AQA June 2003]

[Total: 7]

---

2 Nuclei, like atoms, absorb and emit electromagnetic radiation in the form of photons. For a single nucleus of deuterium (^2_1H), a minimum energy of 2.2 MeV is required to **just** separate all the nucleons.

(a) Explain what is meant by

(i) a nucleon, **[1]**

(ii) a photon. **[1]**

(b) Calculate the wavelength of the electromagnetic radiation absorbed by the nucleus of deuterium to just free all the nucleons.
Data: $1\text{ eV} = 1.6 \times 10^{-19}\text{ J}$
$c = 3.0 \times 10^8\text{ m s}^{-1}$
$h = 6.63 \times 10^{-34}\text{ J s}$ **[3]**

[Total: 5]

Answers on pages 53–57 **Answers** on pages 53–57 **Answers** on pages 53–57

3 The Sun produces energy by means of fusion reactions. One such reaction is

$^{1}H + {}^{1}H \rightarrow {}^{2}H + e^{+} + \nu$

(a) Identify the particles ^{1}H and e^{+}. [2]

(b) Explain how the reaction above releases energy. [2]

(c) Explain why very high temperatures are necessary for the above reaction to occur. [2]

(d) The graph below shows the variation of the average binding energy per nucleon (B.E./nucleon) with the nucleon number *A*.

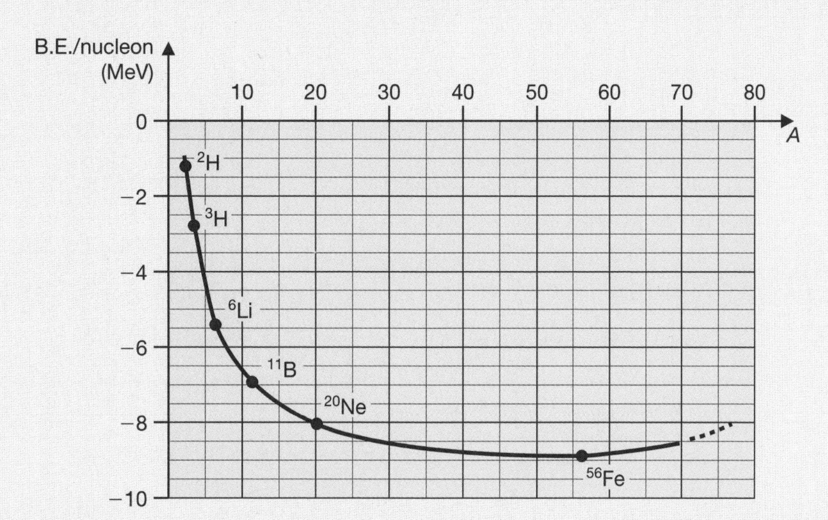

(i) Explain why ^{1}H does not appear on the graph. [1]

(ii) Use the graph to determine the energy released in the reaction.
Data: 1 eV = 1.6×10^{-19} J [2]

[Total: 9]

---

4 One possible fission reaction within a nuclear reactor is shown below.

$^{235}U + {}^{1}n \rightarrow$ [ ] $\rightarrow {}^{96}Rb + {}^{138}Cs + X{}^{1}n$

(a) Complete the reaction above by inserting the appropriate symbols in the box. [1]

(b) Identify the number X of neutrons released in the reaction above. [1]

(c) Calculate the energy released in the above reaction.
Data: mass of ^{235}U nucleus = 3.90×10^{-25} kg
mass of ^{96}Rb nucleus = 1.59×10^{-25} kg
mass of ^{138}Cs nucleus = 2.29×10^{-25} kg
mass of ^{1}n = 1.67×10^{-27} kg
$c = 3.00 \times 10^{8}$ m s^{-1} [4]

(d) Determine the energy released by 1 kg of pure uranium-235 fuel. [3]

(e) State one major disadvantage of using nuclear fuel. [1]

[Total: 10]

Answers on pages 53–57 Answers on pages 53–57 Answers on pages 53–57

5 **(a)** State the **three** assumptions, known collectively as the Cosmological Principle. [3]

(b) State Hubble's law and use it to estimate the age of the universe.
Data: Hubble constant, $H_0 \approx 2.1 \times 10^{-18}\ \text{s}^{-1}$ [4]

(c) Explain what is meant by an 'open' universe. [2]

(d) For a 'closed' universe, the mean density of matter within the universe must be greater than $1.9 \times 10^{-26}\ \text{kg m}^{-3}$. Current experimental evidence suggests that the mean density of matter in the universe is about three protons in every cubic metre of space. Discuss what consequence this may have on the fate of the universe.
Data: mass of proton $= 1.7 \times 10^{-27}$ kg. [3]

[Total: 12]

---

6 **(a)** What evidence is there that the universe is expanding? [2]

(b) The diagram shows the position of a spectral line from hydrogen as observed in the laboratory on the Earth and that from a star in a distant galaxy.

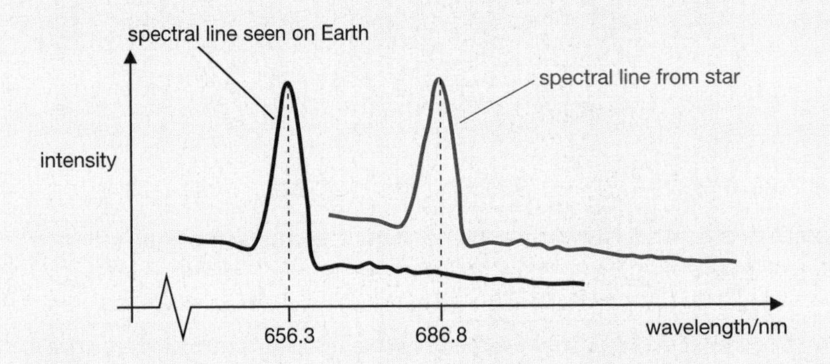

(i) State why the wavelength of light from the star is different from that observed on the Earth. [1]

(ii) For a spectral line of wavelength λ from a stationary source, the change in wavelength $\Delta\lambda$ when the source is moving at a speed v is given by the Doppler equation

$$\frac{\Delta\lambda}{\lambda} = \frac{v}{c}$$

where c is the speed of light in a vacuum.
Calculate the speed of the star as observed from the Earth.
Data: $c = 3.00 \times 10^{8}\ \text{m s}^{-1}$ [2]

[Total: 5]

Particle physics, cosmology/astrophysics and medical physics

 Answers on pages 53–57 Answers on pages 53–57 Answers on pages 53–57

7 This question is about our Solar System and Kepler's third law.

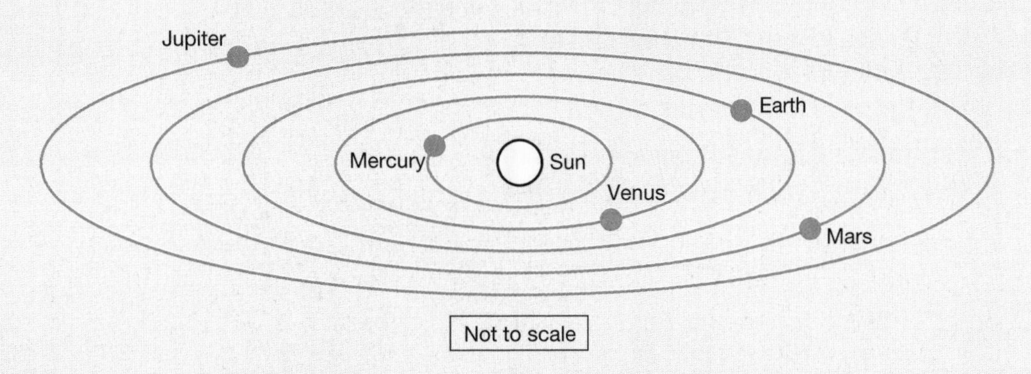

Not to scale

(a) Show that the period T of a planet orbiting our Sun at a mean distance r away from it is given by

$$T^2 = \frac{4\pi^2}{GM} r^3$$

where M is the mass of the Sun. The equation given above shows that $T^2 \propto r^3$, which is an expression for Kepler's third law. **[3]**

(b) The table below shows information about some of the planets in our Solar System.

Planet	T / years	r / A.U.
Mercury	0.24	0.39
Venus	0.61	0.72
Earth	1.00	1.00
Mars	1.88	1.52
Jupiter	11.86	5.20

1 A.U. = mean distance between the Sun and the Earth.

Use the information given in the table to show that Kepler's third law applies to the planets in our Solar System. **[3]**

(c) Our Sun, together with its Solar System, is moving through space because our galaxy is rotating. The mean distance of our Sun from the galactic centre is 25 000 light-years and its orbital period is 2.0×10^8 years.
Use the equation given in **(a)** to calculate the mass M contained within the orbit of the Sun and the galactic centre.
Data: $G = 6.67 \times 10^{-11}$ N m^2 kg^{-2}
$c = 3.0 \times 10^8$ m s^{-1} **[3]**

[Total: 9]

8 **(a)** On the axes below, sketch the variation with frequency f of the minimum intensity I of sound detectable by a person with normal hearing.
Useful data: Ear is most sensitive at 4 kHz.
Threshold intensity, $f_0 \approx 1 \times 10^{-12}\,\mathrm{W\,m^{-2}}$

[3]

(b) In a factory, Health Officers record sound intensity to be $4.1 \times 10^{-2}\,\mathrm{W\,m^{-2}}$.
It is recommended that workers in the factory wear ear guards so as to reduce the sound incident on the ears to $3.0 \times 10^{-5}\,\mathrm{W\,m^{-2}}$.
Calculate the reduction in sound intensity, in dB. [4]

[Total: 7]

---

9 **(a)** Describe some of the effects of ionising radiation on human beings. [2]

(b) Radioisotopes have many uses in non-invasive techniques of diagnosis.
Discuss the use of a specific tracer in medicine. [5]

[Total: 7]

Answers on pages 53–57 Answers on pages 53–57 Answers on pages 53–57

Answers

(1) (a) (i) positron and neutrino

(ii) electron, proton and negative muon

(b) (i) $\mu^- \Rightarrow e^- + \nu_\mu + \bar{\nu}_e$

(ii) Difference: Electrons are stable but muons decay with a very short lifetime.
Similarity: Both are leptons or both have a charge of magnitude e.

---

(2) (a) (i) nucleon → a proton or a neutron.

(ii) photon → quantum (or packet) of electromagnetic radiation.

(b) E = energy of photon

$E = 2.2\ \text{MeV} = 2.2 \times 10^6 \times 1.6 \times 10^{-19} = 3.52 \times 10^{-13}\ \text{J}$

$E = hf$ and $c = f\lambda$

$$E = \frac{hc}{\lambda}$$

$$\lambda = \frac{6.63 \times 10^{-34} \times 3.0 \times 10^8}{3.52 \times 10^{-13}}$$

$\lambda = 5.65 \times 10^{-13} \approx 5.7 \times 10^{-13}\ \text{m}$

examiner's tip

The photons are energetic enough to split the nucleus. The photons belong to the γ-ray region of the electromagnetic spectrum. The other method for separating the nucleons would be to use energetic particles in the form of electrons and protons.

---

(3) (a) ^{1}H is a proton.
e^+ is a positron (the anti-particle of the electron).

(b) In the reaction, mass **decreases**.
According to Einstein's equation, $\Delta E = \Delta mc^2$, a decrease in mass means that energy is released in the reaction.

examiner's tip

Some of the mass in this reaction is converted into energy. This is in accordance with Einstein's equation mentioned above.

(c) The protons are positively charged, they therefore **repel** each other.
At high temperature, the protons move faster and therefore the chance of overcoming the repulsive force is greater.

examiner's tip

The protons may be assumed to behave like the molecules of an ideal gas. Therefore, the mean kinetic energy E_k of the protons is given by

$$E_k = \tfrac{3}{2}kT$$

As the temperature increases, E_k increases. For the protons to 'fuse' together, they must be close enough to be influenced by the strong nuclear force. This force is very short-ranged. The kinetic energy of the protons must be greater than the electrical potential energy between the protons. This happens when the temperature is about 10^8 K.

(d) (i) ^{1}H is a proton.

It has no binding energy because there are no other nucleons.

(ii) B.E. / nucleon ≈ -1.2 MeV

E = energy **released**

$E \approx 2 \times (1.2 \times 10^{6} \times 1.6 \times 10^{-19})$

$E \approx 3.8 \times 10^{-13}$ J

---

(4) (a) ^{236}U label inserted within the box.

examiner's tip

The new nucleus must be an isotope of uranium with an extra neutron. The extra neutron within the nucleus of uranium-236 makes it unstable. After a short period of time, it decays.

(b) The nucleon number must be conserved.

There are two neutrons released in the reaction.

(c) Δm = change in mass

$\Delta m = (1.59 \times 10^{-25} + 2.29 \times 10^{-25} + 2 \times 1.67 \times 10^{-27})$
$\quad -(3.90 \times 10^{-25} + 1.67 \times 10^{-27})$

$\Delta m = -3.30 \times 10^{-28}$ kg

$\Delta E = \Delta mc^2$

$\Delta E = -3.30 \times 10^{-28} \times (3.0 \times 10^{8})^2$

$\Delta E = -2.97 \times 10^{-11}$ J

examiner's tip

The change in mass is negative which implies there is a reduction in mass. The mass is converted into energy, as given by Einstein's equation. The energy is released as kinetic energy of all the particles produced in the reaction.

(d) $\text{Number of nuclei} = \dfrac{1\text{ kg}}{\text{mass of uranium-235 nucleus}}$

$\text{Number of nuclei} = \dfrac{1}{3.90 \times 10^{-25}} = 2.5641 \times 10^{24}$

Energy released $= 2.5641 \times 10^{24} \times 2.97 \times 10^{-11}$

Energy released $= 7.615 \times 10^{13}$ J $\approx 7.62 \times 10^{13}$ J

(e) Radioactive waste from nuclear fuel remains active for a very long time.

(5) (a) The matter is spread evenly (homogeneous universe).
The universe is the same in all directions (isotropic universe).
The laws of physics apply to all inertial frames.

(b) recession speed of galaxy $\propto$ distance of galaxy from us

$v = H_0 d$

$$\text{age of universe} = \frac{1}{H_0}$$

$$\text{age of universe} = \frac{1}{2.1 \times 10^{-18}} \approx 4.8 \times 10^{17}\,\text{s} \qquad (15 \times 10^{19}\ \text{years})$$

examiner's tip

In defining Hubble's law, the reference to a galaxy is vital. Many candidates in exams often state that 'speed $\propto$ distance' and therefore would lose at least one mark.

(c) An open universe will expand forever because the actual mean density of the universe is less than the critical density.

(d) current density $\approx 3 \times 1.7 \times 10^{-27} = 5.1 \times 10^{-27}\,\text{kg m}^{-3}$
This density is less than $1.9 \times 10^{-26}\,\text{kg m}^{-3}$, hence the universe will be 'open', with the universe expanding forever.
However, the existence of dark matter (which is difficult to detect) could make the actual density greater than the current figure.

---

(6) (a) All distant galaxies are moving away from us.
Light from such sources is red-shifted.

examiner's tip

Red-shift means that all the wavelengths of light from the moving source are longer than if the source was stationary.

(b) (i) The star is receding from us, therefore the light from it is red-shifted.

(ii) $$\frac{\Delta\lambda}{\lambda} = \frac{(686.8 - 656.3)}{656.3}$$

$$\frac{\Delta\lambda}{\lambda} = 0.04647$$

$v = 0.04647 \times 3.00 \times 10^{8}$
$v \approx 1.39 \times 10^{7}\,\text{m s}^{-1}$

Particle physics, cosmology/astrophysics and medical physics

(7) (a) $F = \frac{GMm}{r^2}$ $\quad F = m(\omega^2 r)$ and $\omega = \frac{2\pi}{T}$

Therefore $\frac{GM}{r^2} = \omega^2 r$

$$\frac{GM}{r^3} = \frac{4\pi^2}{T^2}$$

$$T^2 = \frac{4\pi^2}{GM} r^3$$

examiner's tip

This proof is worth remembering. It uses all the key ideas from Newtonian mechanics and Newton's law of gravitation.

(b) If $T^2 \propto r^3$, then $\frac{T^2}{r^3}$ must be a constant.

The values for $\frac{T^2}{r^3}$ are

0.971, 0.997, 1.000, 1.006 and 1.000
for Mercury, Venus, Earth, Mars and Jupiter respectively.

Hence Kepler's third law is applicable to our Solar System.

(c) $T^2 = \frac{4\pi^2}{GM} r^3$

where $r = 25\,000 \times 3.0 \times 10^8 \times 365 \times 24 \times 3600 = 2.365 \times 10^{20}$ m
and $T = 2.0 \times 10^8 \times 365 \times 24 \times 3600 = 6.307 \times 10^{15}$ s

$$M = \frac{4\pi^2 r^3}{GT^2}$$

$$M = \frac{4\pi^2 \times (2.365 \times 10^{20})^3}{6.67 \times 10^{-11} \times (6.307 \times 10^{15})^2}$$

$M = 1.96 \times 10^{41}$ kg $\approx 2.0 \times 10^{41}$ kg (Equivalent to ~ 10^{11} solar masses)

examiner's tip

It is vital that the distance and the period are calculated in metres and seconds respectively. If this is not done, then most likely you would lose marks since the final mass will not be in kilograms. This is an 'extension' question where a specific model is applied to another situation. In this case, Kepler's third law which has been shown to hold true for the planets in our Solar System is now being applied to the entire galaxy.

(8) (a) The minimum ought to have the following values:

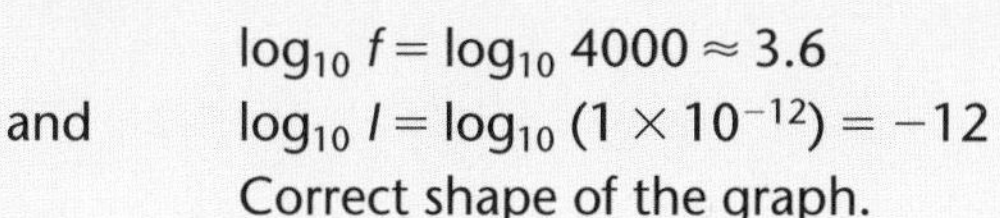

$\log_{10} f = \log_{10} 4000 \approx 3.6$

and $\log_{10} I = \log_{10} (1 \times 10^{-12}) = -12$

Correct shape of the graph.

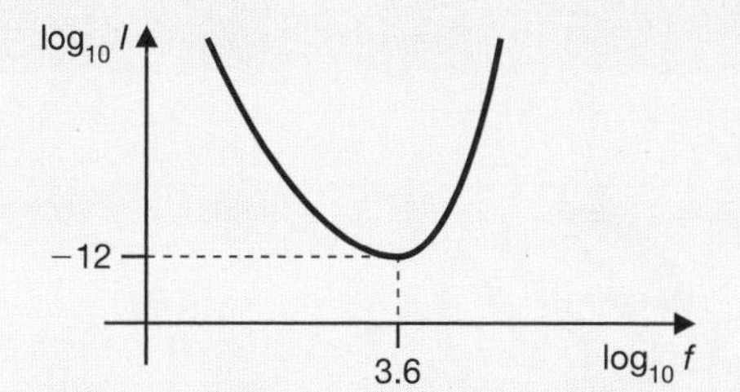

(b) Intensity level $= 10 \log_{10} \left(\frac{I}{I_0}\right)$

Original value $= 10 \log_{10} \left(\frac{4.1 \times 10^{-2}}{10^{-12}}\right) = 106.1$ dB

Final value $= 10 \log_{10} \left(\frac{3.0 \times 10^{-5}}{10^{-12}}\right) = 74.8$ dB

Reduction $= 106.1 - 74.8 = 31.3$ dB

Reduction ≈ 31 dB.

---

(9) (a) Any *one* from the first two statements plus third statement:

Genetic mutation of cells

Cells may be destroyed

leading to serious illness or death.

(b) A tracer is a radioactive substance introduced into the body of the patient. It is used for detecting abnormalities in the function of the organs. A commonly used tracer is ^{99}Tc, an isotope of **technetium** which emits γ-rays and has a short half-life of 6 hours. The half-life is short enough for little damage to be done to living cells. ^{99}Tc is used for detection of tumours.

examiner's tip

The chances are that the examiners may also assess the candidate's 'Quality of Written Communication' in a question like (b). It is therefore very important that extra care is given to the construction of sentences.

A2 Mock Exam 1

Centre number ____________

Candidate number ____________

Surname and initials ____________

Examining Group

Physics

Time: 1 hour Maximum marks: 60

Instructions

Answer **all** questions in the spaces provided. Show all steps in your working.
The marks allocated for each question are shown in brackets.
Any data required for a question are given where appropriate.

Grading

Boundary for A grade	48/60
Boundary for C grade	36/60

---

1 (a) ***Momentum*** is a ***vector*** quantity.
Explain the meanings of the terms in italics.

....

.... [3]

(b) Explain what is meant by an elastic collision.

....

.... [2]

(c) Gas atoms in the Earth's atmosphere constantly collide with each other. In one such collision, a hydrogen atom of mass 1.7×10^{-27} kg travelling at 420 $\mathrm{m\,s^{-1}}$, makes a head-on collision with a stationary oxygen atom of mass 2.7×10^{-26} kg. After the impact, the oxygen atom moves with a speed v and the hydrogen atom rebounds with a speed 370 $\mathrm{m\,s^{-1}}$. This collision event is shown in the diagram below.

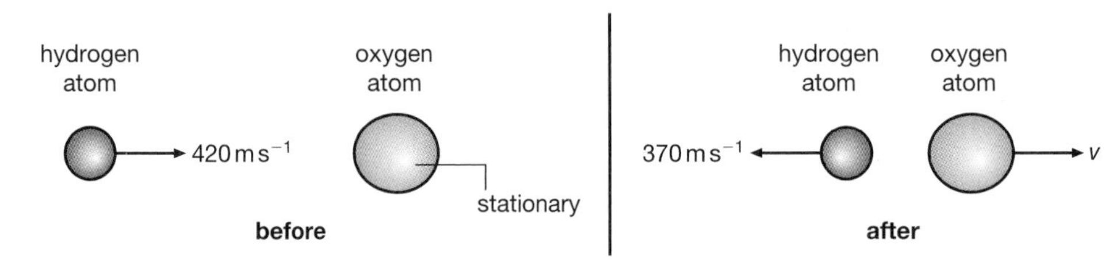

(i) State what is conserved in all collisions.

.... [1]

(ii) Calculate the magnitude of the velocity v of the oxygen atom after the collision.

....

.... [3]

[9 marks]

2 A cyclist is travelling on a banked circular track at a constant speed of $12\ \text{m s}^{-1}$. The cyclist is travelling in a horizontal circle of radius 40 m and is at right angles to the track as shown in the diagram.

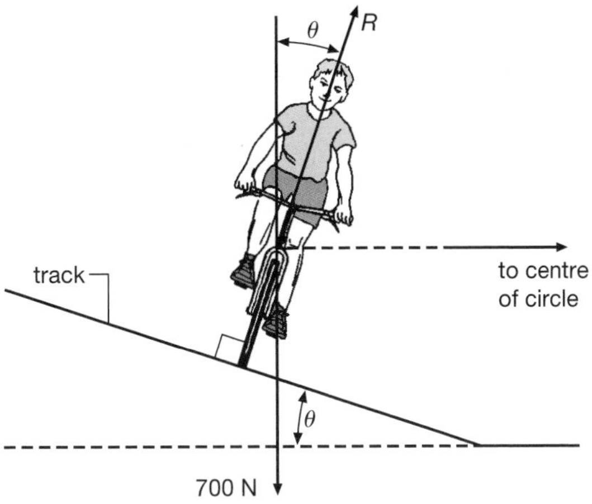

The combined weight of the cyclist and the bicycle is 700 N. The track exerts a contact force R on the cyclist. This force is at right angles to the track.

(a) For the combination of the cyclist and the bicycle, calculate

(i) the centripetal acceleration,

....

....

.... [2]

(ii) the centripetal force.
Data: $g = 9.8\ \text{N kg}^{-1}$

....

....

.... [3]

(b) Determine the angle θ made by the track with the horizontal.

....

....

.... [4]

[9 marks]

3 The diagram shows the path of a proton travelling in a vacuum at 5.0×10^7 m s^{-1} as it enters and leaves a region of uniform magnetic field of flux density 1.4 T. The direction of the field is into the plane of the paper.

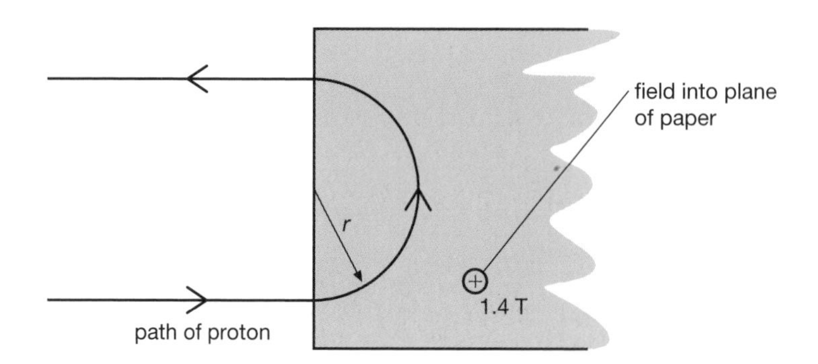

(a) Explain why the speed of the proton is not affected by the presence of the magnetic field.

....

.... [2]

(b) The proton describes a circular path in the region of the magnetic field.
For the proton, calculate

(i) its acceleration,
Data: mass of proton $= 1.7 \times 10^{-27}$ kg
$e = 1.6 \times 10^{-19}$ C

....

.... [3]

(ii) the radius of the circular path in the field.

....

.... [2]

(c) Show that the time spent in the magnetic field by the proton is independent of its speed v.

....

.... [3]

[10 marks]

4 (a) State Faraday's law of electromagnetic induction.

....

.... [1]

(b) The diagram shows a circular coil of area 3.5×10^{-4} m^2 and having 1000 turns placed in a uniform magnetic field having a flux density B of 0.75 T.
The plane of the coil is at right angles to the direction of the magnetic field.

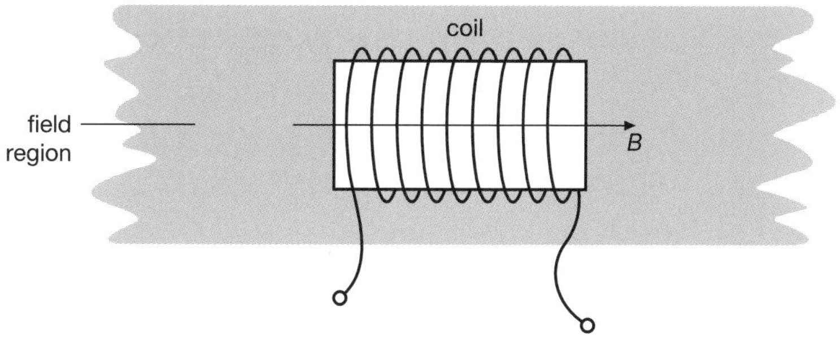

coil

field region

B

(i) State why there is no e.m.f. induced across the ends of the coil when the coil is moved at a constant speed in a direction parallel to the magnetic field lines.

....

.... [1]

(ii) Calculate the flux linkage for **each** turn of the coil.

....

....

.... [3]

(iii) The coil is removed from the magnetic field in a time of 20 ms. Calculate the magnitude of the mean e.m.f. induced across the ends of the coil.

....

....

.... [3]

(iv) State another way in which an e.m.f. may be induced across the ends of the coil.

....

.... [1]

[9 marks]

5 **(a)** Explain what is meant by the ***binding energy*** of a nucleus.

....

....

....

.... **[1]**

(b) The graph below shows the average binding energy per nucleon (B.E./nucleon) against the nucleon number *A*.

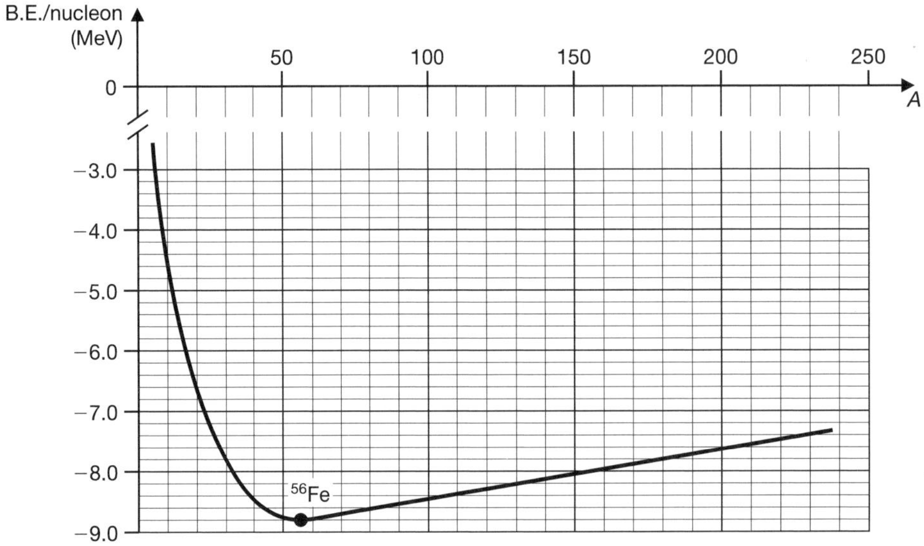

B.E./nucleon (MeV)

0, −3.0, −4.0, −5.0, −6.0, −7.0, −8.0, −9.0

50, 100, 150, 200, 250 *A*

^{56}Fe

(i) What is the significance of the average binding energy per nucleon being negative?

....

....

.... **[1]**

(ii) Use the graph to determine the binding energy of a single nucleus of ^{56}Fe.

....

....

.... **[2]**

(c) One of the naturally occurring isotopes of radium is radium-226. The nuclei of radium-226 emit alpha (α) particles.

(i) Complete the following nuclear decay equation for a single nucleus of radium-226.

$$^{226}_{88}Ra \rightarrow Rn + He$$ [2]

(ii) Calculate the energy released by the decay of a single nucleus of radium-226.

Data: $1\,u = 1.66 \times 10^{-27}$ kg

$c = 3.00 \times 10^{8}$ ms^{-1}

mass of alpha-particle = 4.00 u

mass of Ra nucleus = 225.98 u

mass of Rn nucleus = 221.97 u

....

....

....

.... [4]

(iii) State in what form the energy in **(c)(ii)** is released.

....

....

.... [1]

(iv) The half-life of radium-226 is about 1600 years. Suggest one reason why radium cannot be used as a viable source of energy.

....

....

.... [1]

[12 marks]

6 (a) Define simple harmonic motion.

....

.... **[2]**

(b) A metal sphere of mass 230 g is attached to the end of a spring. The metal sphere is pushed up and then released. The graph below shows the variation of the position of the sphere with time *t*.

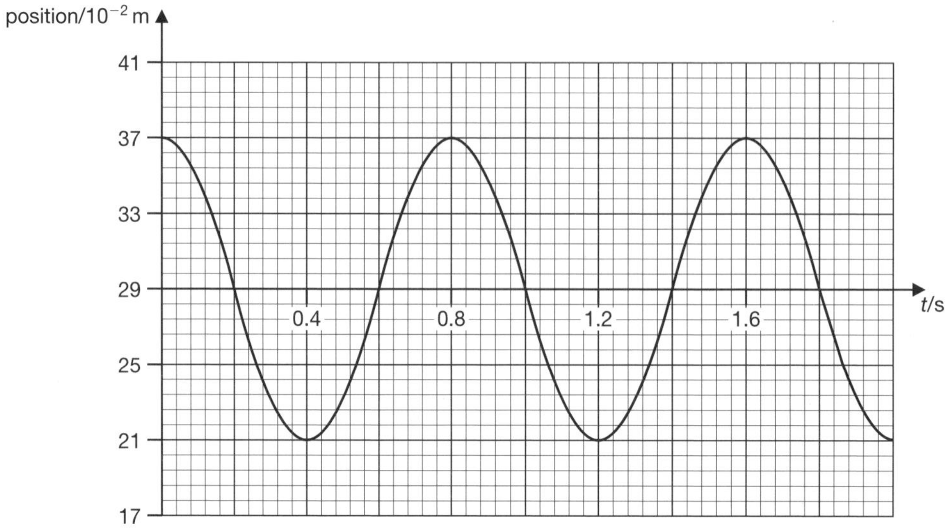

position/10^{-2} m

41, 37, 33, 29, 25, 21, 17

0.4, 0.8, 1.2, 1.6

t/s

(i) State the amplitude of the oscillation.

....

.... **[1]**

(ii) State the times at which the potential energy of the oscillating system is a minimum. Explain your answer.

....

.... **[2]**

(iii) Calculate the maximum net force acting on the metal sphere.

....

....

.... **[3]**

(iv) Calculate the position of the metal sphere after 2.5 s.

....

....

.... **[3]**

[11 marks]

A2 Mock Exam 2

Centre number ____________________
Candidate number ____________________
Surname and initials ____________________

Examining Group

Physics

Time: 1 hour Maximum marks: 60

Instructions

Answer **all** questions in the spaces provided. Show all steps in your working.
The marks allocated for each question are shown in brackets.
Any data required for a question are given where appropriate.

Grading
Boundary for A grade 48/60
Boundary for C grade 36/60

---

1 A proton enters a region of uniform magnetic field strength of flux density 90 mT. The proton is initially travelling at right angles to the magnetic field. In the magnetic field, the proton describes an arc of a circle of radius 28 cm. The diagram shows the proton about to enter the magnetic field.

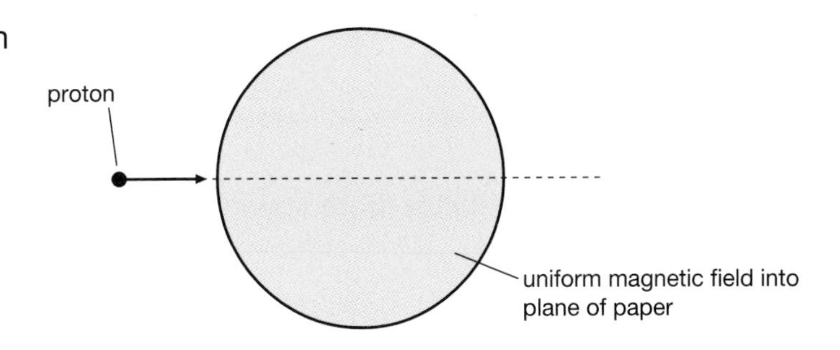

Data: mass of proton $= 1.7 \times 10^{-27}$ kg
$e = 1.6 \times 10^{-19}$ C

(a) Complete the diagram above to show the path described by the proton as it travels through the magnetic field region. [2]

(b) For this proton,

(i) show that its speed is 2.4×10^{6} m s^{-1},

....

.... [3]

(ii) calculate its momentum,

....

....

.... [2]

(iii) calculate the centripetal force experienced.

....

.... [2]

[9 marks]

2 A 1000 μF capacitor is connected to a d.c. supply of e.m.f. 24 V and having an internal resistance of 5.0 Ω.

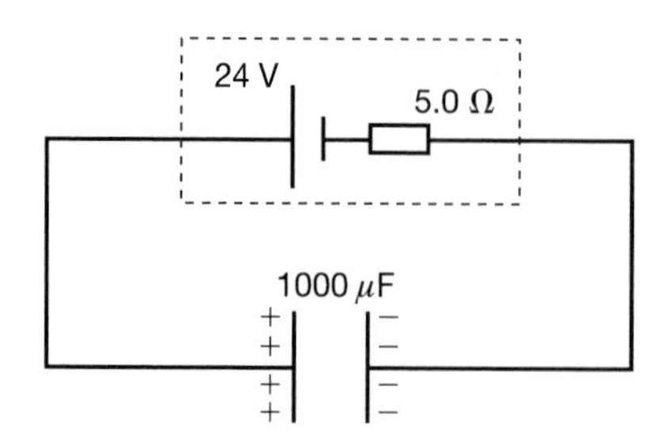

(a) Explain why the charge stored by the capacitor does not depend on the internal resistance of the d.c. supply.

....

....

....

....

.... [1]

(b) Calculate the energy stored by the capacitor.

....

....

....

....

....

.... [2]

(c) The fully charged capacitor is disconnected from the d.c. supply and placed within an electronic circuit that is designed to utilise the energy stored by the capacitor to produce a short burst of visible light. The circuit draws a steady current from the capacitor for duration of 1.2 ms and converts all the stored energy of the capacitor into visible light.

(i) Calculate the current drawn from the capacitor.

....

....

....

.... **[2]**

(ii) Calculate the light power produced by the electronic circuit.

....

....

....

.... **[2]**

(iii) Estimate the total number of photons of light emitted when the capacitor is discharged by the circuit.

Data: average wavelength of visible light $= 5.5 \times 10^{-7}$ m

$h = 6.6 \times 10^{-34}$ J s

$c = 3.0 \times 10^{8}$ m s^{-1}

....

....

....

.... **[3]**

[10 marks]

3 **(a)** The diagram below shows the track of an alpha-particle travelling close to the nucleus of a gold atom.

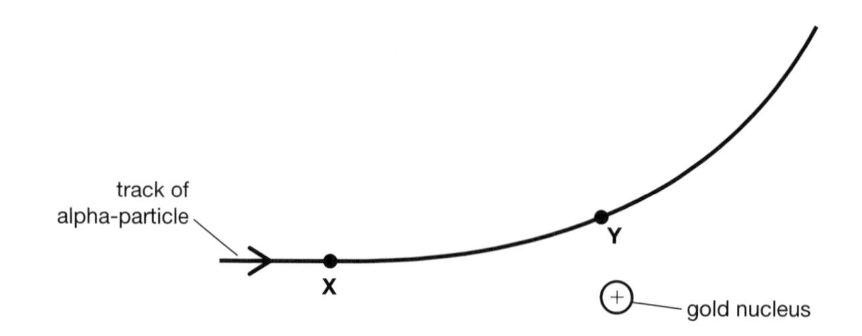

(i) On the diagram above, draw two arrows, to show the directions and relative magnitudes of the forces experienced by the alpha-particle when at the points **X** and **Y**. [2]

(ii) Describe what happens to the energy of the alpha-particle as it moves closer to and then moves further away from the gold nucleus.

....

....

....

....

....

....

....

....

....

.... [3]

(b) The diagram below shows an arrangement to demonstrate the electrical repulsion between two charged metal spheres.

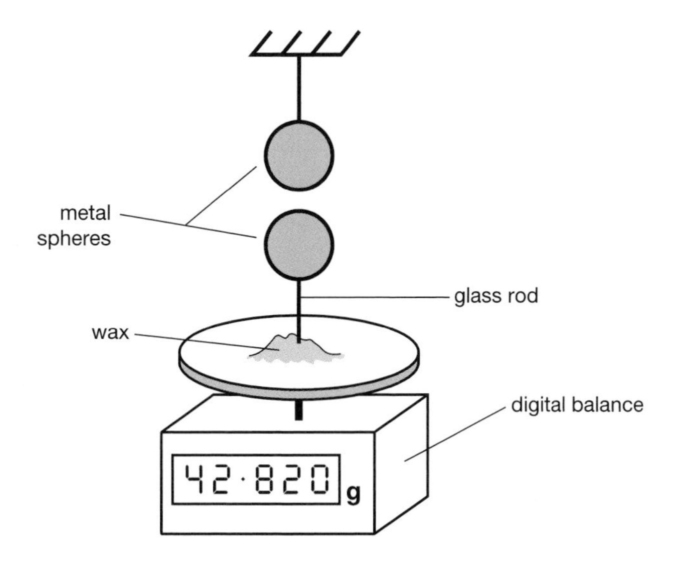

The separation between the centres of the spheres is 4.5 cm. Both metal spheres are initially uncharged. The balance reading is 42.820 g. Both metal spheres are given an equal amount of charge by momentarily touching the spheres to the positive electrode of an e.h.t. supply. The separation between the spheres remains constant and the new balance reading is 42.880 g.

(i) Use the information given above to calculate the charge on each metal sphere.

Data: $g = 9.8\ \text{N kg}^{-1}$

$\varepsilon_0 = 8.85 \times 10^{-12}\ \text{F m}^{-1}$

....

....

....

.... **[3]**

(ii) The radius of each metal sphere is 2.0 cm. Use your answer to **(b)(i)** to calculate the electrical potential at the surface of each sphere.

....

....

....

.... **[3]**

[11 marks]

4 (a) Define the gravitational field strength at a point in space.

....

....

....

.... [1]

(b) (i) Show that the gravitational field strength at a distance r from a point mass M is given by

$$g = -\frac{GM}{r^2}$$

where G is the gravitational constant $6.67 \times 10^{-11}\ N\ m^2\ kg^{-2}$

[2]

(ii) Calculate the mass of the planet Mercury given that its surface gravitational field strength is $3.8\ N\ kg^{-1}$ and its radius is 2.4×10^6 m.

....

....

....

....

....

.... [3]

(c) One of the most famous 'thought experiments' was Newton's cannon-ball experiment. The diagram below shows a cannon-ball fired at right angles to the gravitational field of the Earth. The speed of the cannon-ball is such that it does not land or escape the Earth, but instead, describes a circle equal to the radius of the Earth.

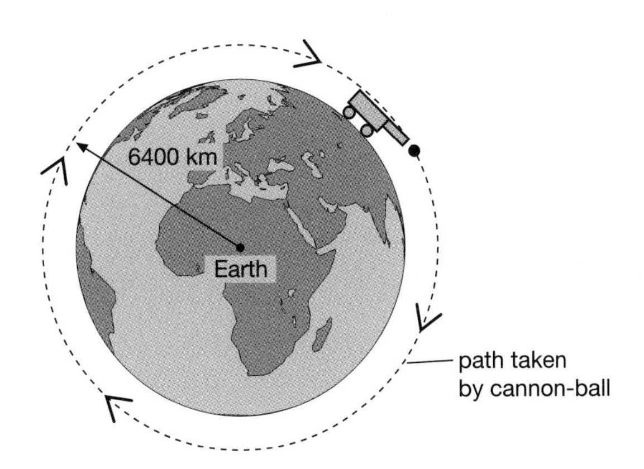

Data: $g = 9.8\ \text{N kg}^{-1}$

radius of the Earth = 6400 km

(i) State the centripetal acceleration of the cannon-ball.

....

....

.... **[1]**

(ii) Calculate the time it would take for the cannon-ball to describe one complete orbit.

....

....

....

....

....

.... **[3]**

[10 marks]

5 **(a)** The mean radius of the ^{4}He nucleus is 1.9×10^{-15} m. Calculate the mean density of the helium-4 nucleus. You may assume that a neutron has the same mass as a proton.

Data: mass of proton $\approx 1.7 \times 10^{-27}$ kg

....

....

.... [3]

(b) Explain why the mean density of any nucleus is independent of its mass.

....

....

.... [3]

(c) The mean separation of the protons within the nucleus of helium-4 is about 3.8×10^{-15} m.

(i) Calculate the gravitational force, F_G between the two protons.
Data: $G = 6.67 \times 10^{-11}$ N m^2 kg^{-2}

....

....

.... [2]

(ii) Calculate the electrical force, F_E between the two protons.
Data: $\varepsilon_0 = 8.85 \times 10^{-12}$ F m^{-1}
$e = 1.6 \times 10^{-19}$ C

....

....

.... [2]

(iii) Explain why the answers to **(c)(i)** and **(c)(ii)** suggest that there must be another force acting on the protons when inside the nucleus.

....

....

.... [2]

[12 marks]

6 (a) In 1961, Jönsson carried out experiments that provided further evidence that electrons were diffracted by very narrow slits. This is illustrated in the diagram below.

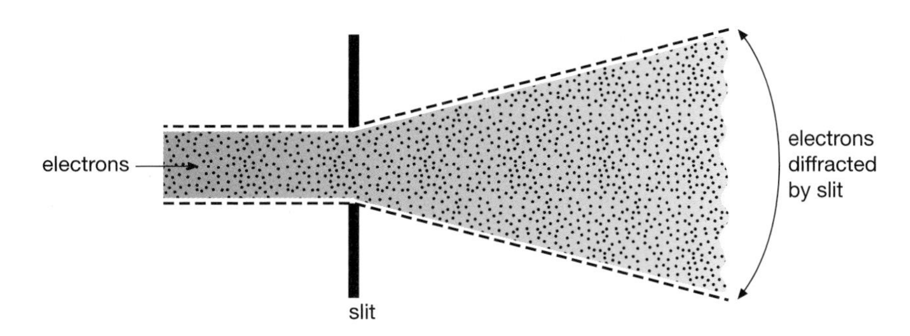

State what may be interpreted about the nature of electrons from such diffraction experiments.

....

.... **[1]**

(b) In a small television tube, electrons travel at a speed of $2.6 \times 10^{7}\,\mathrm{m\,s^{-1}}$. For an electron, calculate

(i) its momentum,

Data: mass of electron = 9.1×10^{-31} kg

....

.... **[2]**

(ii) the de Broglie wavelength λ.

Data: $h = 6.63 \times 10^{-34}$ J s

....

....

.... **[3]**

(c) Explain why a person of mass 65 kg running at $9.0\,\mathrm{m\,s^{-1}}$ through an open door will not exhibit diffraction effects.

....

....

.... **[2]**

[8 marks]

A2 Synoptic Paper

Centre number ____________________
Candidate number ____________________
Surname and initials ____________________

Letts Examining Group

Physics

Time: 1 hour Maximum marks: 60

Instructions
Answer **all** questions in the spaces provided. Show all steps in your working.
The marks allocated for each question are shown in brackets.
Any data required for a question are given where appropriate.

Grading
Boundary for A grade 48/60
Boundary for C grade 36/60

---

1 There are several conservation laws in physics. Name **three** conservation laws and give two examples for each conservation law from your study of A-level physics.

....

....

....

....

....

....

....

....

....

....

....

....

....

....

.... [9]

[9 marks]

2 **(a)** The charge Q on a capacitor discharging through a resistor is given by the equation

$$Q = Q_0 e^{-\frac{t}{CR}}$$

(i) Define the following terms:

Q_0 ....

CR .... [2]

(ii) State and explain the unit for the product CR.

....

....

.... [2]

(iii) In terms of CR, determine the time taken for the charge on the capacitor to halve.

....

....

....

.... [3]

(b) The decay of radioactive nuclei may also be described by an equation similar to that for the decay of charge on a capacitor as it discharges through a resistor. Write an equation for the number N of active nuclei left in a sample at a particular time t. Describe one similarity and one difference between the decay of nuclei and the decay of charge on a discharging capacitor.

....

....

....

....

....

....

.... [3]

[10 marks]

3 The universe is believed to have originated from a Big Bang some 1.8×10^{10} years ago. At the early stages of the universe, photons were constantly being converted into particles and vice versa. When a particle interacts with its antiparticle, they completely annihilate each other. The mass of the particles is converted into two photons of electromagnetic radiation.

One such event was the interaction between the electron and its antiparticle the positron. The interaction may be summarised as follows:

$$e^{+} + e^{-} \Rightarrow \gamma + \gamma$$

(a) Calculate the total energy released when an electron and positron annihilate each other.

Data: $c = 3.0 \times 10^{8}\ \mathrm{m\,s^{-1}}$

mass of electron $= 9.1 \times 10^{-31}$ kg

....

....

....

.... [3]

(b) Calculate the wavelength of the electromagnetic radiation released in the form of the photons.

Data: $h = 6.6 \times 10^{-34}$ J s

....

....

....

.... [3]

(c) Calculate the momentum of an electron if it were to have a de Broglie wavelength equal to that calculated in **(b)**.

....

....

....

.... [3]

[9 marks]

4 According to the wave-particle duality, **particles** like electrons can exhibit 'wave-like' behaviour and electromagnetic **waves** can exhibit a 'particle-like' behaviour.

(a) For an **electron**, outline the evidence available for

(i) it being a particle,

....

....

....

.... [3]

(ii) it having a 'wave-like' behaviour.

....

....

....

.... [3]

(b) For an electromagnetic **wave**, outline the evidence available for

(i) it having a 'particle-like' behaviour,

....

....

....

.... [3]

(ii) it being a wave.

....

....

....

.... [3]

[12 marks]

5 Two sources of renewable energy are solar energy and wind energy.

In a wind turbine, the kinetic energy of the wind is used to generate electrical energy. The maximum power P developed by a wind turbine is given by

$$P = \tfrac{1}{2}\rho A v^2$$

where ρ is the density of air, A the area swept by the turbine blades and v is the speed of the wind. A typical wind turbine in a windy region of the country produces an output power of 20 kW when the average wind speed is $8.0\ \text{m s}^{-1}$.

A solar panel transforms the incident light into electrical energy. A particular solar panel, of cross-sectional area $1.2\ \text{m}^2$, produces a current of 7.2 A and an electromotive force (e.m.f.) of 18 V when light of intensity $900\ \text{W m}^{-2}$ is incident normally on the panel.

(a) Explain what is meant by a renewable source.

....

....

.... [1]

(b) Determine the output power from a wind turbine in a region with an average wind speed of $10\ \text{m s}^{-1}$.

....

....

.... [2]

(c) For the solar panel, calculate

(i) its output electrical power,

....

....

.... [3]

(ii) its efficiency.

....

....

.... [2]

(d) Name **one** possible reason for the solar panel not being 100% efficient.

.... [1]

(e) Calculate the total surface area of the solar panels needed so that its output electrical power is equal to the power output of 20 kW from a single wind turbine.

....

....

....

....

.... [3]

(f) Suggest how you may overcome the problem of energy requirements, after sunset, of a small remote community that is reliant on solar panels.

....

....

....

....

.... [2]

[14 marks]

6 Some physicists would strongly argue that physics is the study of various **fields** and how they interact with matter. Describe **three** different types of fields. In each case, define the field strength and state the unit of the field strength.

....

....

....

....

....

....

....

....

....

....

....

....

....

....

....

....

....

....

....

....

....

.... [6]

[6 marks]

A2 Mock Exam 1 Answers

(1) (a) momentum = mv

where m is the mass of the object and v is its **velocity**.

A vector quantity has both magnitude **and** direction.

(b) In an elastic collision both momentum and kinetic energy are conserved.

examiner's tip

In all collisions, energy is conserved. However, in your response you must make it clear that it is the kinetic energy that is conserved in an elastic collision. When the kinetic energy is converted into other forms, such as heat, sound etc., the collision is then referred to as an inelastic collision.

(c) (i) Momentum is conserved in all collisions.

(ii) Total initial momentum = Total final momentum

$1.7 \times 10^{-27} \times 420 = (2.7 \times 10^{-26} \times v) + (1.7 \times 10^{-27} \times -370)$

$v = (7.14 \times 10^{-25} + 6.29 \times 10^{-25})/2.7 \times 10^{-26}$

$v = 49.7 \approx 50\ \text{m s}^{-1}$

examiner's tip

Momentum is a vector quantity. After the collision, the hydrogen atom is moving in the opposite direction. Its momentum must therefore be assigned an opposite sign. To get the answer, you must first set up the equation in accordance with the principle of conservation of momentum and then proceed to solve it in terms of *v*.

(2) (a) (i) $a = \dfrac{v^2}{r}$

$a = \dfrac{12^2}{40}$

$a = 3.6\ \text{m s}^{-2}$

(ii) $m = \dfrac{\text{weight}}{g}$

$m = \dfrac{700}{9.8} = 71.4\ \text{kg}$

$F = ma$

$F = 71.4 \times 3.6$

$F = 257 \approx 260\ \text{N}$

(b) Horizontally $\Rightarrow R \sin \theta = 257$

Vertically $\Rightarrow R \cos \theta = 700$

Therefore $\tan \theta = \dfrac{257}{700}$

$\theta = 20.2° \approx 20°$

examiner's tip

The net force in the horizontal direction is responsible for the circular motion of the cyclist. There is no net force in the vertical direction. These two statements provide the necessary physics for this question. The question also requires a knowledge of the following piece of mathematics:

$$\tan\theta = \frac{\sin\theta}{\cos\theta}$$

(3) (a) The force on the proton is **perpendicular** to the velocity.
The speed is the same since no work is done by this force on the proton.

(b) (i) $F = Bqv$

$F = 1.4 \times 1.6 \times 10^{-19} \times 5.0 \times 10^{7}$

$F = 1.12 \times 10^{-11}$ N

$$a = \frac{F}{m}$$

$$a = \frac{1.12 \times 10^{-11}}{1.7 \times 10^{-27}}$$

$a = 6.59 \times 10^{15} \approx 6.6 \times 10^{15}$ m s^{-2}

(ii) $$a = \frac{v^2}{r}$$

$$r = \frac{(5.0 \times 10^{7})^2}{6.59 \times 10^{15}}$$

$r = 0.379 \approx 0.38$ m

(c) $F = ma$

$$Bqv = m\left(\frac{v^2}{r}\right)$$

$$\left(\frac{v}{r}\right) = \frac{Bq}{m}$$

$$\text{time} = \frac{\text{distance}}{\text{speed}}$$

$$\text{time} = \frac{\pi r}{v}$$

$$\text{time} = \pi\left(\frac{m}{Bq}\right)$$

The time spent in the field is independent of the speed v.

(4) (a) The e.m.f. induced in a circuit $\propto$ rate of change of flux linkage.

(b) (i) There is no change in the flux linking the coil.

examiner's tip

For an e.m.f. to be induced across the ends of the coil, there must be a change in the magnetic flux linkage. Flux linkage is equal to *NBA*. None of these quantities is changing with respect to time.

(ii) flux linkage = NBA

For one turn, flux linkage = ϕ $\quad (N = 1)$

$\phi = BA$

$\phi = 0.75 \times 3.5 \times 10^{-4}$

$\phi = 2.63 \times 10^{-4} \approx 2.6 \times 10^{-4}$ Wb

examiner's tip

The magnetic flux linking each turn is given by

$$\phi = BA \cos \theta$$

where θ is the angle between the normal of the coil and the magnetic field. Since the magnetic field is normal to the plane of the coil, the angle θ is equal to 0° and therefore the cos θ factor is equal to unity.

(iii) e.m.f. = − rate of change of flux linkage

change in flux linkage = $1000 \times 2.63 \times 10^{-4}$ Wb

$$\text{e.m.f.} = \frac{1000 \times 2.63 \times 10^{-4}}{20 \times 10^{-3}}$$

e.m.f. = 13.2 ≈ 13 V

(iv) Rotate the coil in the magnetic field.

examiner's tip

The magnetic flux linking each turn of the coil is given by

$$\phi = BA \cos \theta$$

where θ is the angle between the normal of the coil and the magnetic field. By changing the angle θ, the flux linking each turn may be altered with respect to time and therefore an e.m.f. induced across the ends of the coil.

(5) (a) The binding energy of a nucleus is numerically equal to the energy required to completely separate the nucleus into its individual nucleons.

(b) (i) The B.E. per nucleon is **negative** because it implies that external energy must be supplied to the nucleus in order to separate the nucleons.

examiner's tip

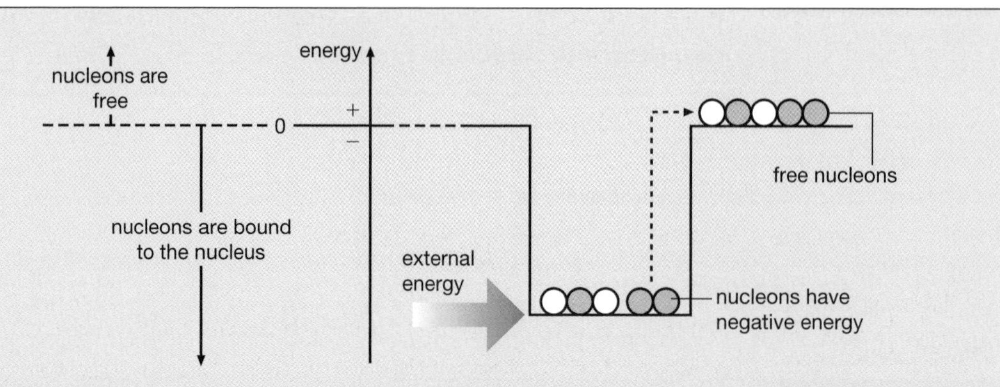

Within the nucleus, the nucleons are held together by the *strong nuclear force.* To move the nucleons apart requires external energy. Hence the initial energy of the nucleus must be negative. If the energy was assigned a positive sign, then this would imply that the nucleons have kinetic energy and are already free from the effect of the strong interaction.

(ii) B.E. = number of nucleons × B.E. /nucleon

B.E. ≈ 56 × (−8.8 MeV)

B.E. ≈ −490 MeV

(c) (i) ${}^{226}_{88}\text{Ra} \rightarrow {}^{222}_{86}\text{Rn} + {}^{4}_{2}\text{He}$

Correct nucleon and proton numbers shown for the α-particle and the daughter nucleus of radon Rn.

(ii) Change in mass = Δm

$\Delta m = (221.97 + 4.00) - 225.98$

$\Delta m = -\ 0.01$ u

$\Delta m = -\ 0.01 \times 1.66 \times 10^{-27} = -\ 1.66 \times 10^{-29}$ kg

$\Delta E\ = \Delta mc^2$

$\Delta E\ = \Delta mc^2 = 1.66 \times 10^{-29} \times (3.00 \times 10^8)^2$

$\Delta E\ = 1.494 \times 10^{-12} \approx 1.49 \times 10^{-12}$ J

(iii) The energy is released as the kinetic energy of the α-particle and the Rn nucleus.

examiner's tip

In this question, the change in mass is identified as the energy released as kinetic energy of the by-products in the decay. According to Einstein's equation,

$$\Delta E = \Delta mc^2$$

a *decrease* in mass means that energy is *released* in the decay. An increase in mass would imply that external energy is required for some event to occur. For an everyday event like walking, the change in our body mass is insignificant and we are therefore oblivious of it. For nuclear events, however, a tiny change in mass becomes extremely significant especially when there are a very large number of nuclei present.

(iv) The energy is released over a long period of time and, therefore, the power of the source would be very small.

(6) (a) The acceleration of the object is directly proportional to its displacement from the equilibrium position. The acceleration is always directed towards the equilibrium position.

(b) (i) amplitude = 37 − 29 = 8.0 cm

(ii) From the position against time graph, the **gradient** is equal to velocity. The sphere has maximum speed and hence kinetic energy, when it travels through the equilibrium position. The total energy of the sphere remains constant.

(iii) $\omega = \frac{2\pi}{T} = \frac{2\pi}{0.8} = 7.855 \text{ rad s}^{-1} \approx 7.9 \text{ rad s}^{-1}$

$F = ma = m(\omega^2 A)$

$F = 0.230 \times (7.855)^2 \times 0.08 = 1.135 \approx 1.14 \text{ N}$

examiner's tip

The acceleration is a maximum when displacement is equal to the amplitude. Do not forget to convert the amplitude into metres.

(iv) displacement, $x = A\cos(\omega t)$

$x = 8.0 \cos(7.855 \times 2.5) \approx 5.6 \text{ cm}$

position = 29 + 5.6 = 34.6 cm

examiner's tip

The answer requires the position of the object and not its displacement. You therefore have to add the 29 cm to the calculated displacement. Also, do not forget that the curve given in the question is a cosine graph. Your calculation must be 'radian' mode when you do this calculation.

A2 Mock Exam 2 Answers

(1) (a)

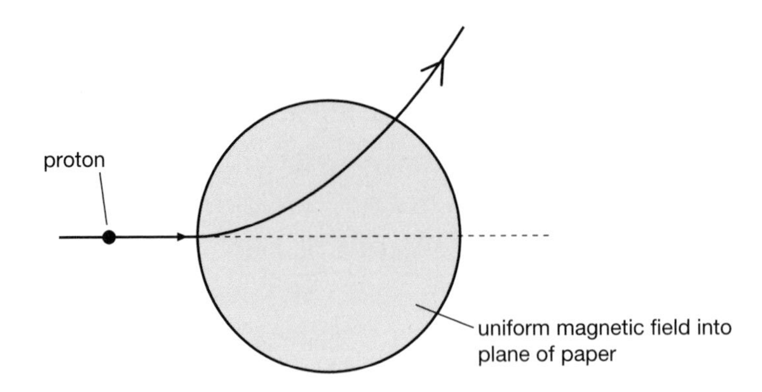

(b) (i) $BQv = \frac{mv^2}{r}$

$BQ = \frac{mv}{r}$

$v = \frac{BQr}{m} = \frac{0.090 \times 1.6 \times 10^{-19} \times 0.28}{1.7 \times 10^{-27}} = 2.372 \times 10^6 \text{ m s}^{-1}$

$v \approx 2.4 \times 10^6 \text{ m s}^{-1}$

examiner's tip

This is a tough question that requires an understanding of both how moving particles interact with a magnetic field and circular motion. You cannot determine the speed v of the particle without first deriving the equation

$$v = \frac{BQr}{m}.$$

(ii) $p = mv$

$p = 1.7 \times 10^{-27} \times 2.372 \times 10^6 \approx 4.0 \times 10^{-21} \text{ kg m s}^{-1}$

(iii) $F = BQv$

$F = 0.090 \times 1.6 \times 10^{-19} \times 2.372 \times 10^6 \approx 3.4 \times 10^{-14} \text{ N}$

examiner's tip

You can also get the answer by using

$$F = \frac{mv^2}{r}.$$

(2) (a) The capacitor will stop charging when the p.d. across it is equal to the e.m.f. of the supply. The final p.d. across the capacitor, and hence the charge it stores, is independent of the resistance in the circuit.

examiner's tip

The resistance in the circuit, together with the capacitance, determines the rate at which the capacitor charges. The final charge will still be determined by the final p.d. across the capacitor which is 24 V.

(b) $E = \frac{1}{2}V^2C$

$E = \frac{1}{2} \times 24^2 \times 1000 \times 10^{-6} = 0.288\ \text{J}$

examiner's tip

You can also determine the energy stored by using the equations

$$Q = VC$$

followed by

$$E = \frac{1}{2}QV$$

The equation above is an easier route.

(c) (i) $I = \dfrac{Q}{t}$

$$I = \frac{VC}{t} = \frac{24 \times 1000 \times 10^{-6}}{1.2 \times 10^{-3}} = 20\ \text{A}$$

(ii) power $= \dfrac{\text{energy}}{\text{time}}$

$$\text{power} = \frac{0.288}{1.2 \times 10^{-3}} \approx 240\ \text{W}$$

(iii) energy of photon $= \dfrac{hc}{\lambda}$

$$= \frac{6.6 \times 10^{-34} \times 3.0 \times 10^{8}}{5.5 \times 10^{-7}} = 3.6 \times 10^{-19}\ \text{J}$$

$$\text{number of photons} = \frac{0.288}{3.6 \times 10^{-19}} = 8.0 \times 10^{17}$$

examiner's tip

In this question, you have to assume that all the energy stored by the capacitor is used to produce the photons. You must therefore calculate the energy of each photon of light.

(3) (a) (i)

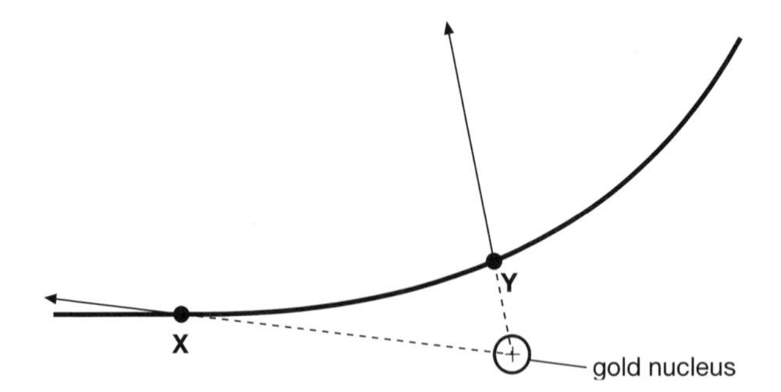

examiner's tip

The force acting on the alpha-particle is directly away from the gold nucleus. The arrow at Y is longer because the force experienced by the alpha-particle is greater $\left(F \propto \frac{1}{r^2}\right)$.

(ii) As the alpha-particle gets closer to the nucleus, there is a loss in kinetic energy and gain in potential energy. At the distance of closest approach, the kinetic energy is a minimum (but not zero, because the collision is not a 'head-on' collision). As the alpha-particle recedes away from the positive nucleus, its kinetic energy increases. When the alpha-particle is far away from the nucleus it has no potential energy.

(b) (i) $F = (42.880 - 42.820) \times 10^{-3} \times 9.8 = 5.88 \times 10^{-4}$ N

$$F = \frac{Qq}{4\pi\varepsilon_0 r^2} \qquad (Q = q)$$

$$F = \frac{Q^2}{4\pi\varepsilon_0 r^2}$$

$$Q = \sqrt{4\pi\varepsilon_0 r^2 F} = \sqrt{4\pi \times 8.85 \times 10^{-12} \times 0.045^2 \times 5.88 \times 10^{-4}}$$

$$Q = 1.15 \times 10^{-8}\,\text{C}$$

examiner's tip

In this question, it is vital that you change the two balance readings into the force experienced by the sphere resting on the balance. There is an increase in the balance reading from 42.820 g to 42.880 g because of the repulsion between the two positively charged metal spheres. The force on each sphere is 5.88×10^{-4} N.

(ii) $V = \dfrac{Q}{4\pi\varepsilon_0 r}$

$$V = \frac{1.15 \times 10^{-8}}{4\pi \times 8.85 \times 10^{-12} \times 0.02}$$

$$V = 5.2\,\text{kV}$$

(4) (a) gravitational field strength = force per unit mass

examiner's tip

You can write $g = \frac{F}{m}$, as long as you define all the symbols in the equation.

(b) (i) $F = -\frac{GMm}{r^2}$ Equation for Newton's law of gravitation

Therefore $\frac{F}{m} = -\frac{GM}{r^2}$

But the gravitational field strength is $g = \frac{F}{m}$.

Hence $g = -\frac{GM}{r^2}$

examiner's tip

In order to show the correct answer, you must annotate or explain all steps of your proof.

(ii) $g = -\frac{GM}{r^2}$ Magnitude only

$$M = \frac{gr^2}{G} = \frac{3.8 \times (2.4 \times 10^6)^2}{6.67 \times 10^{-11}} \approx 3.3 \times 10^{23}\ \text{kg}$$

(c) (i) The acceleration must be equal to the acceleration of free fall, namely $9.8\ \text{m s}^{-2}$.

examiner's tip

It is vital to appreciate that the centripetal acceleration of the cannon-ball is equal to the acceleration of free-fall at the surface of the Earth. Therefore, the acceleration of the cannon-ball in its circular orbit is equal to $9.8\ \text{m s}^{-2}$.

(ii) $a = \frac{v^2}{r}$

$$v = \sqrt{ar} = \sqrt{9.8 \times 6400 \times 10^3} = 7.92 \times 10^3\ \text{m s}^{-1}$$

$$\text{time taken} = \frac{\text{distance}}{\text{speed}}$$

$$\text{time taken} = \frac{2\pi r}{v} = \frac{2\pi \times 6400 \times 10^3}{7.92 \times 10^3} \approx 5.1 \times 10^3\ \text{s}$$ (1.4 hours or 85 minutes)

(5) (a) $\rho = \dfrac{M}{V}$

$M = 4 \times 1.7 \times 10^{-27} = 6.8 \times 10^{-27}$ kg

$V = \frac{4}{3}\pi r^3 = \frac{4}{3}\pi \times (1.9 \times 10^{-15})^3$

$V = 2.873 \times 10^{-44}\ \text{m}^3$

$\rho = \dfrac{6.8 \times 10^{-27}}{2.873 \times 10^{-44}}$

$\therefore\ \rho = 2.37 \times 10^{17} \approx 2.4 \times 10^{17}\ \text{kg m}^{-3}$

(b) Mass $\propto A$

$r = r_o A^{1/3}$ and $V = \frac{4}{3}\pi r^3$

$\therefore$ volume $\propto A$

density $= \dfrac{\text{mass}}{\text{volume}}$

The density is independent of the nucleon number A.

examiner's tip **The mean density of the nucleus is the same because it consists of the same matter.**

(c) (i) $F_G = \dfrac{GMm}{r^2}$ (Magnitude only)

$F_G = \dfrac{6.67 \times 10^{-11} \times (1.7 \times 10^{-27})^2}{(3.8 \times 10^{-15})^2}$

$F_G = 1.33 \times 10^{-35} \approx 1.3 \times 10^{-35}$ N

(ii) $F_E = \dfrac{Qq}{4\pi\varepsilon_0 r^2}$

$F_E = \dfrac{(1.6 \times 10^{-19})^2}{4\pi\varepsilon_0(3.8 \times 10^{-15})^2}$

$F_E = 15.9 \approx 16$ N

(iii) The gravitational force is weak but attractive and the electrical force is much larger but repulsive. With the forces above, there is a **net repulsive** force. There must therefore be another attractive force holding the protons.

examiner's tip **The nucleons are held within the nucleus by the strong nuclear interaction. This interaction exists between the quarks that make up the nucleons.**

(6) (a) **Moving** electrons behave like **waves.**

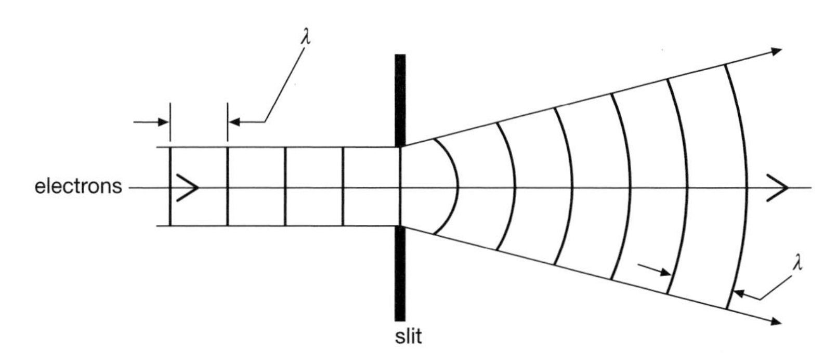

examiner's tip

The waves associated with moving particles are not electromagnetic in nature. They have their own peculiar characteristics. The waves are referred to as either 'de-Broglie waves' or 'matter waves'.

(b) (i) $p = mv$

$p = 9.1 \times 10^{-31} \times 2.6 \times 10^{7}$

$p = 2.37 \times 10^{-23} \approx 2.4 \times 10^{-23}\ \text{kg m s}^{-1}$

(ii) $\lambda = \dfrac{h}{p}$

$$\lambda = \frac{6.63 \times 10^{-34}}{2.37 \times 10^{-23}}$$

$\lambda = 2.80 \times 10^{-11} \approx 2.8 \times 10^{-11}\ \text{m}$

(c) The wavelength of the person is too small (compared with the width of the door).

$$\lambda = \frac{6.63 \times 10^{-34}}{(65 \times 9.0)} \sim 10^{-36}\ \text{m}$$

examiner's tip

For diffraction effects to be significant, the wavelength λ must be comparable to the size of the 'gap'. For a person to show diffraction effects, the gap size ought to be about 10^{-36} m, which is an impossibility. Electrons may be diffracted by matter because their de-Broglie wavelength can be comparable to the separation between the atoms.

A2 Synoptic Paper Answers

SYNOPTIC PAPER

(1) Conservation of momentum

- The decay of radioactive nuclei. For example, a uranium nucleus decaying into an alpha-particle and a thorium nucleus. The momentum of the alpha-particle is equal in magnitude to that of the thorium nucleus but in the opposite direction.
- A trolley colliding with another trolley on a linear air-track. The total initial momentum of the trolleys is equal to the final momentum of the trolleys after impact.

examiner's tip

This is an open-ended question. You may gain marks by other relevant responses like:
'In a collision, momentum is always conserved, but the kinetic energy may not be conserved. Such a collision is referred to as an inelastic collision'.

Conservation of charge

- When two deuterium nuclei fuse together to form helium-4, charge is conserved. The initial charge is $+2e$ and final charge is also $+2e$.

  $${}^{2}_{1}\text{H} + {}^{2}_{1}\text{H} \Rightarrow {}^{4}_{2}\text{He}$$

- According to Kirchhoff's first law, the sum of currents entering a point is equal to the sum of the current out of the point. This is because charge is conserved.

Conservation of energy

- When a car brakes, its initial kinetic energy is transformed into heat and sound.
- In a filament lamp, the input electrical energy is equal to the sum of light and heat output.

examiner's tip

There are other quantities conserved. These are listed below:
- **Nucleon and proton numbers when a nucleus decays**
- **Mass and energy (Einstein's equation: $\Delta E = \Delta mc^2$)**

(2) (a) (i) Q_0 is the initial charge on the capacitor.
CR is the time constant of the circuit.

(ii) The product *CR* has unit seconds.

$-\frac{t}{CR}$ have no units. Since time *t* is measured in seconds, *CR* must also be measured in seconds.

examiner's tip

The base of natural logs, e, must have an exponent that has no units. Hence $\frac{-t}{CR}$ has no unit, which in turn implies that *t* and *CR* must have the same unit.

(iii) $Q = Q_0 e^{-\frac{t}{CR}}$

Since $Q = \frac{Q_0}{2}$, we have $\frac{1}{2} = e^{-\frac{t}{CR}}$

$2 = e^{\frac{t}{CR}}$

Taking logs to the base *e*, we have:

$\ln 2 = \frac{t}{CR}$

Therefore
$t = CR \ln 2 \propto 0.693\, CR$

(b) The equation for radioactive decay of nuclei: $N = N_0 e^{-\lambda t}$
Both the number *N* and charge *Q* decay **exponentially** with respect to time.
Radioactive decay is a random event and shows 'statistical scatter' in the graph of *N* against *t*.

examiner's tip

The sketches below illustrate the last point made above. The Q against t graph is a smooth curve, whereas N against t shows 'small bumps' due to the random decay of the nuclei.

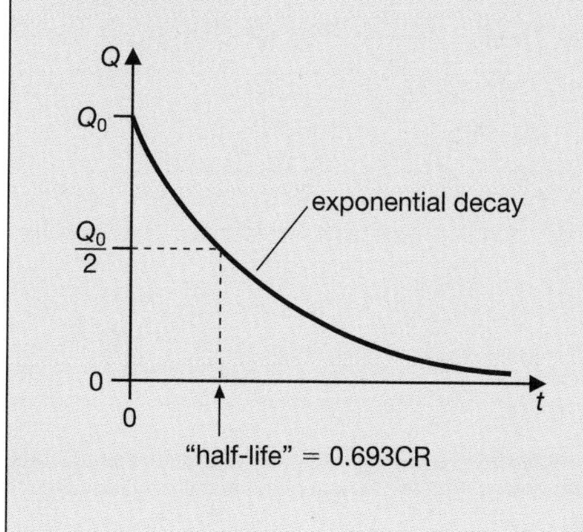

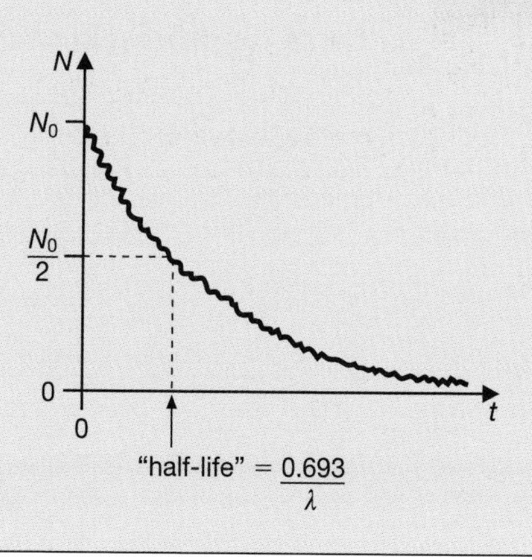

(3) (a) $\Delta E = \Delta mc^2$

energy released $= 2 \times 9.1 \times 10^{-31} \times (3.0 \times 10^8)^2$

$= 1.638 \times 10^{-13}\,\text{J} \approx 1.64 \times 10^{-13}\,\text{J}$

examiner's tip

The mass of the positron is the same as that for the electron. Hence, the total mass change is equal to twice the mass of the electron. There is a decrease in mass, hence according to Einstein's equation, energy is released.

(b) $E = \frac{1}{2} \times 1.638 \times 10^{-13} = 8.19 \times 10^{-14}\,\text{J}$ E = energy of each photon

$$E = \frac{hc}{\lambda}$$

$$\lambda = \frac{hc}{E} = \frac{6.6 \times 10^{-34} \times 3.0 \times 10^8}{8.19 \times 10^{-14}}$$

$\lambda = 2.42 \times 10^{-12} \approx 2.4 \times 10^{-12}\,\text{m}$

examiner's tip

Do not forget that there are two photons emitted. Hence the energy of each γ-ray photon is $8.19 \times 10^{-14}\,\text{J}$

(c) $\lambda = \dfrac{h}{p}$

$$p = \frac{6.6 \times 10^{-34}}{2.42 \times 10^{-12}}$$

$p \approx 2.7 \times 10^{-22}\,\text{kg m s}^{-1}$

examiner's tip

It is vital that de Broglie waves are not confused with electromagnetic waves. Some candidates in exams, think that $\lambda = \dfrac{h}{mc}$. This of course, is not the case.

(4) (a) (i) An electron has mass and charge. It can be deflected by both electric and magnetic fields. A moving electron has momentum. When electrons collide with other particles, momentum is conserved.

(ii) An electron travels through space as a wave. The wavelength λ is given by the de Broglie equation, $\lambda = \frac{h}{p}$ where h is the Planck constant and p is the momentum of the electron. Electrons can be diffracted by either atoms or atomic nuclei. Electrons passing through two-slits can also show 'interference' effects.

(b) (i) The interaction of electromagnetic waves with matter is explained using the model of photons. A photon is a quantum of electromagnetic radiation. The energy of a photon is given by $E = hf$, where h is the Planck constant and f is the frequency of the radiation. In order to explain the photoelectric effect, a single photon interacts with a single electron on the surface of the metal and energy is conserved in this interaction.

(ii) Electromagnetic waves passing through a narrow gap show diffraction effects. With multiple slits, these waves can also show interference effects. Electromagnetic waves also refract when travelling into another medium.

examiner's tip

There are many other points you can put down, but the ones outlined above are the main points related to the wave-particle duality.

(5) (a) The energy supply will not run out. It is constantly being replaced.

(b) $P \propto v^2$

$P = 20 \times \left(\frac{10}{8}\right)^2 = 31.3\text{ kW}$

(c) (i) $P = VI$

$P = 18 \times 7.2 = 129.6 \approx 130\text{ W}$

(ii) input power $= 1.2 \times 900 = 1080$ W

output power $= 129.6$ W

efficiency $= \frac{129.6}{1080} \times 100 = 12\%$

(d) The incident light is reflected off the solar panels.

(e) A 1.2 m^2 solar panel gives an output power of 129.6 W.

Therefore

total surface area $= \frac{20\,000}{129.6} \times 1.2 = 185.2 \approx 190\text{ m}^2$

(f) The electrical energy could be stored in capacitors during the day.

At night time, the energy stored by the capacitors may be used.

(6) Gravitational field

Gravitational field strength g is defined as the force per unit mass $\left(g = \frac{F}{m}\right)$.

Unit: N kg^{-1}

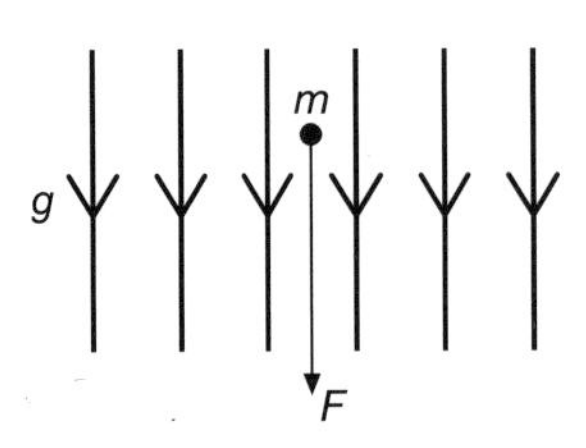

Magnetic field

Magnetic flux density B is defined as $B = \frac{F}{IL}$, where F is the force experienced by a conductor of length L carrying a current I placed at right angles to the magnetic field.

Unit: tesla (T)

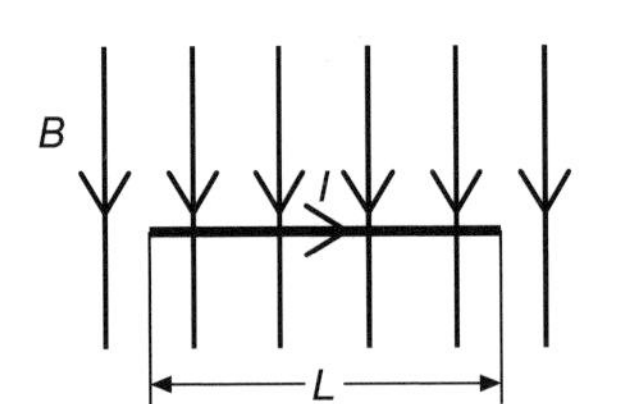

The force F is into the plane of the paper, as given by Fleming's left-hand rule.

Electric field

Electric field strength is defined as the force experienced per unit positive charge $\left(E = \frac{F}{Q}\right)$

Unit: N C^{-1}